THE
UNREALISTS

⤙ WILLIAM JAMES ⤚
He yearned toward mystery

THE
UNREALISTS

JAMES, BERGSON, SANTAYANA, EINSTEIN,
BERTRAND RUSSELL, JOHN DEWEY,
ALEXANDER AND WHITEHEAD

BY

HARVEY WICKHAM

Essay Index Reprint Series

BOOKS FOR LIBRARIES PRESS
FREEPORT, NEW YORK

80m Essay 2nd.

First Published 1930
Reprinted 1970

STANDARD BOOK NUMBER:
8369-1736-7

LIBRARY OF CONGRESS CATALOG CARD NUMBER:
78-105051

PRINTED IN THE UNITED STATES OF AMERICA

DEDICATED

TO THOSE WHO WONDER AT WHAT
PRESIDENT HOOVER HAS CALLED "THE
SUBSIDENCE OF OUR FOUNDATIONS"

CONTENTS

LIST OF ILLUSTRATIONS

ix

THE
UNREALISTS

CHAPTER I

WHAT DO YOU KNOW

I. THE OUT THERE

YOU know very little, and possibly I know less, yet we know rather more than sometimes we are inclined to admit. I know, for instance, that I exist. Let us assume that you know as much. It is exhilarating, even appalling, the amount of knowledge that this implies.

Permit me to speak of it in the first person. I was once totally unfamiliar with life. It was a new thing. I lay, so they tell me, in a cradle. I was able—so they tell me again—to react in a vague sort of way to some of the things about me. But I did not know this, so I have some doubt as to what extent it was really I.

And then, one day—or it may have been one night—I saw a light. Saw it not with my eyes only, but *saw* it. I became aware of something which was not myself, something *out there!* And at that same instant I knew that I was.

I remember this. I do not claim to remember the first time it happened, or that the experience was altogether sudden and complete. It is not complete yet. There remain depths within and without that no one has fathomed. But it happens to a certain extent every morning, every time I wake from a doze over some philosopher who tells me I am never awake and sends me to sleep forthwith by way of proof. Even while I sleep, something wakes. If I forget, something remembers, giving me a sense of continuity. Consciousness is born again and again—not out of total unconsciousness, perhaps, but total enough so far as my own consciousness is concerned. The point I wish to make, however, is that one becomes aware of Self only by becoming aware of the Not-Self. Singly and alone, we do not exist.

Now how did I know that the light out there was not myself? Simply because it came, it moved, it went, in obedience to some will not my own. How did I know it was not a part of my body? I didn't. My body was out there, too. It had what I have since learned to call weight, and offered considerable resistance to my efforts to bring it into subjection. I reached with my hands for this and that, but for a long time my aim was very bad. Other things were yet more difficult to manage, so I came to distinguish my body from the bedclothes.

Generally the bedclothes seemed friendly, but not always. They would sometimes tie themselves into knots, and at other times slide off to the floor. They were capable of bringing Smothering Heat, or of leaving me exposed to Cold—two implacable enemies of my ally, Comfortable Warmth. Then there was Floor, good when it permitted itself to be crept upon, bad when it flew up and struck me in the face. It was not above running splinters into my hands and knees.

I became a fierce partisan of my body, for whatever hurt it hurt me. It was moved at times to hurt me on its own account, more particularly through its gastro-enteric region, and I eventually learned that its very submission was specious and full of treachery. So from the first it was evident that this was a world of sweets and sours. Not even parents were consistently agreeable.

Pleasure came when things yielded to my will, pain when they put up a fight—glorious hazard, bearing in its hands the delight of victory or the crush of defeat. My will was unlimited in its range of demands, but its ability to compel exterior obedience was lamentably small. Not but what the universe bowed down to a very considerable extent, but it was by no means enough. There were irresistible forces against which no physical headway could be made. For me, the dining-table refused to budge, and if I beat it in the leg the only noticeable result was a bruised fist.

Did I therefore conclude that the leg was bad in a moral sense? No; it was its evil disposition that I blamed, for I personified Leg, endowing it with a will, a soul as well as body. That things had wills I could not doubt. I knew how I asserted myself, and saw no reason for attempting the impossible task of inventing a different and totally unimaginable motive-power for anything else. Will was Cause.

Yet I could never come into direct contact with Cause. It was always Body that I met on my strolls. I was handicapped, I suppose, by the fact of having a body of my own. It was impossible even to think of a bodiless ego. Saint Paul, I notice, suffered from a like defect. No sooner did he leave off speaking of a body whose glory was terrestrial than he began to speak of one whose glory should be celestial. Even ghosts have bodies, of a sort, if you believe in ghosts. Think what would happen if two wills were to meet with nothing between! They would not be two wills, but only one. There has to be separation to make two, and it is through this separation that we have to act.

But to those unable to think of a soul without a body, the modernist philosopher offers the choice of thinking of a body without a soul—something through which nothing acts. And again I have to acknowledge my incompetence. If nothing acts, nothing reaches me. That which acts becomes will—the only kind of force of which I can even dream, since it is the only kind that I possess. True, I often speak of physical forces—muscular, nervous, electrical, and so on—but I only mean some force that is not mine. If it is not like mine, then I know nothing about it, for I am compelled to go by such knowledge as I have, not by such as I have not. My own body is largely dominated by forces that are not mine. Generally my muscles move at my suggestion, but sometimes they merely twitch. I don't make them do it, or even want them to do it. I want my heart to keep on beating, but I can't make it beat, and someday the will that lies beyond the beating will certainly will it to rest. The forces of dissolution will then take it in hand. What are they? The laws of nature. Formerly these were supposed to have something to do with the will of God. Then it was "discovered" that *they* were independent forces. Later it was discovered that they were not forces, but merely the way things happened. We now are supposed to know that they are the ways things would happen if they should happen to happen that way—something like the Volstead act. And the world was made by them!

Nevertheless, I am inclined to think that my body has a soul, if only a feeble one, for soul is but another name for me. If there is no *you* connected with your body, then you may be in a position to think of its action as uncaused from within, and at the same time as uncaused from without; of effects unaffected; of a cause created by its own effects, or of a Universe of Effects without a Universal Cause. But to me, handicapped as I say by being alive, the Universe seems rather a big smoke not to have a bit of fire, nor can I quite follow those who suggest that the fire was caused by a lot of little smokes getting together.

As a matter of fact, I am still a child, steeped in superstition, and with all this smoke in my eyes I can only see wills acting through means to produce results. Principle, instrument, effect! This tripartite division of experience was so drilled into my head by the first hard knocks and soft caresses of life that I have never been able to forget it. Everybody is a Scholastic at two, and Scholastic everybody remains in practice, though by dint of learning a sufficient number of things which aren't so it is possible to lose one's early grip upon the theory.

But there is no use telling a child that the Out There is the same as himself. Did he will to be weaned? Or that he is the same as it. Is he his own weaning? No hope, either, of making him believe that there is no such thing as cause. Are the things that he does himself lacking in cause? And what of himself? Did he cause that? No; he seems to be an effect there. Some effects are so vast and startling! No baby could have brought *them* about. What did? Who made *me*?

God, some will say, thinking it a good enough answer for a child. Very well. Self and Not-Self—thunder, lightning, sun, moon, and stars, father, mother, the flowers in the garden—all are now accounted for. God fills the bill. He is a sort of human being, of course, since that is the only kind of being with which we have any immediate acquaintance—a grown-up being withal; and, like all other grown-ups, not quite accountable but very much in evidence. A childish notion, is it not? Very. But what have you, who have put away childish things, to offer in its stead? Something more comprehensive, no doubt, something less "mystical," something rational, which can be grasped by the mind—a thing especially useful in these days when it is so generally admitted that we have no minds, when it has been logically demonstrated, time and again, that there is no such thing as logic. Now a thing that can be grasped by the mind is a thing created in the mind, a concept, to wit. So what you substitute for God is a human concept. No doubt it accounts for everything that you have done all by yourself without any help, and not counting the trifling assistance given by the fact that you are you to begin with. It ought to be perfectly sufficient—if the world indeed were created by your conceiving of it.

But the God of childhood is not a concept, He is an anthropomorphic deity. The same thing? Hardly. Anthropomorphic means man with a difference. Such a deity includes everything that can be conceived of and everything that can't—the can't doing most of the work. For though we may realize that God is something more than human, we have no experience of being that more. Experience forever extends, but beyond it and within it Mystery extends also. We blow a bubble only to increase the extent of its contact with the air. What, then, is the use of God? If we make him explicable, he isn't God and doesn't explain life. If we let Him explain life, He becomes Himself inexplicable. As a means of bringing everything down to the level of our understanding, He is of no use whatever. I should never think of mentioning Him if I hadn't been born and did not find myself somewhat puzzled by that trifling fact.

Certum est quia impossibile est; or as Jerry puts it in *Jingling in the Wind,* "Life is very peculiar."

"Compared with what?" said the spider.

That is just it. What is there which is not peculiar and totally impossible? Impossible, that is, for our own unaided selves to have brought about.

Yet the mind itself can prove that beyond its own limits there is Something. The prisoner in a cell can discover that there is something beyond him by his mere inability to get out. He may be blind, or in darkness, but the wall that he beats with his hands is, nevertheless, beyond the hands that beat upon it.

2. THE IN HERE

"A half-truth, like a half-brick, is always more forcible as an agrument than a whole one," says Stephen Leacock in *The Garden of Folly;* "It carries further."

Surely Leacock is the prince of realists. For now that we have admitted the existence of the Out There we have to admit the existence of the In Here, in contrast, and so find ourslves with a very difficult and unwieldy whole brick upon our hands. It is no longer possible to be simple, clever, and striking. We are in the midst of the infinite complexities of actual life.

Not even the In Here is simple. We can never penetrate to the core of it, any more than we can penetrate to the core of the beyond. Our very consciousness seems to float over the depths with never a plunge to the very bottom. True, much that is said to happen to what is now called our unconscious or our subconscious mind really happens merely to the body, which stores experience for us—changes that, because of our preoccupations, we feel only later, as when I overlook a disturbance in the abdomen until suddenly I yelp with pain and the doctor hurries me off to the hospital.

In such case, I think, it may be argued that I didn't have the experience until I had it; that it didn't happen to the real me until I felt it. But there are other cases when the disturbance seems to be in the inner depths, closer to me than is even my conscious self. If there is a consciousness there it is like the consciousness of dreams, a good part of which passes through Lethe before we wake. But my own opinion is that the self hovers between two mysteries, and grows only as it becomes aware of them. It is reflection that holds the mirror up not only to Nature, but to Me. I see myself, but as in a glass, darkly. It is as if there were two powers, one of them given to me to exercise, the other to exercise myself upon, either in the way of defiance or submission. And neither is fully known.

That is the objection to the Will—a word that so few philosophers can tolerate. It is inexplicable, creative, miraculous. Yes, it is completely impossible—impossible for us to have brought about. And yet we are forced to admit the creative. Even mechanists admit it, though they may call it Evolution. Two blades of grass grow where but one grew before. The tree-tops of the forests are alive with monkeys; and suddenly (or gradually, it matters not which), there are men poking inquisitively about upon the ground.

It is no more unreasonable for me to speak of my own creative force, which is so weak, than it is to speak of that creative force which continues to bring not only me but my changing environment about—a force so boundless and yet so conscientiously yielding when I exert myself either to further or to thwart it. What is unreasonable is to deny what we cannot understand, or to speak of it in wilfully contradictory terms—as when we say that without any force at all the results of force take place. The doctrine of Will has also been hampered by the assumption that it must be unlimitedly free if it be free at all, when at the last analysis its freedom consists merely in its ability not to yield—itself.

Aside from thinkers avowedly religious, Émile Meyerson, the great Frenchman (naturalized, for he was born in Poland), alone has in modern times given sufficient recognition to the element of the unknown in our predicament. Herbert Spencer canonized it, but he called it the Unknowable, as if its unknowability were a hard surface, impossible in the least to penetrate. Meyerson admits what he terms *irrationals* as constituent parts of all experience. But he is seventy years old; and in spite of his *L'explication dans les sciences, Identité et realité,* and *La déduction relativiste,* his influence has scarcely been felt. Nor is his word, *irrational,* altogether happy. There is nothing irrational, against reason, in the fact that a limited creature has inadequate notions of the Illimitable. His reason would tell him that even his notions of the superior must, in all reasons, be lacking. Superrational is the proper word. The Unknowable is but the depths of the Unknown. No wall bars our approach. Even to call it the Unknowable admits that we know something of its nature, its ineffability, to wit.

The Will of God, to use a quaint term whose superb adequacy only cant could have dragged in the dust, seems not to have exhausted itself in action. Creation still moves. Nor has my own will become the all it might be. This capacity for continual doing, so obvious on every side, drives us to the belief in that inexhaustible reservoir known as Being as distinguished from Existence. This is the Absolute, or Unrelated, the Uncreate, the Unrevealed. We can say nothing whatever about it except that it is not this, nor that, nor the other thing—for in so far as it is this, that or the other thing it is not absolute but has become related to our experience. I do not know whether more philosophy has destroyed itself in trying to deny the Absolute, or in trying to describe it in positive terms. But in reality both these methods of mental suicide are the same. No man attempts to drag the Absolute down to his own level unless he wishes to befuddle himself with the idea that there is no Absolute, absolutely.

But life would be of small consequence if it consisted only in knowing. No sooner, however, do we discover the existence both of ourselves and of something else, than we discover our feelings to be involved. But what are feelings? What, for example, is pain?

It is a wave of a curious kind of chemical decomposition passing through nerve filaments—at least that is all that reveals itself to an outsider deaf to our complaints and of a vivisecting turn of mind. But that is precisely what pleasure is. Even our own sensations and emotions are merely felt. That quality which we call pleasurable or painful is given them by ourselves. The words agreeable and disagreeable express the situation exactly. Any happening of which we are aware, to which the will does not consent, which is contrary to our "inclination," with which we do not agree—is painful. And nothing else is.

So we translate feelings into pains and pleasures,

usually in accordance with an instinct handed down through countless generations which leads us to associate hurts with harm. A pain, from the standpoint of the In Here, is a danger-signal indicating a threat to the continued existence of the body. I do not wish my body to be discontinued. In so far as pain involves the future, it is fear.

It might not be altogether wise to lose the knack of reading these ancient signs in the ancient way. A man who did not know a pain when he felt one might sit tranquilly on a hot stove until he was burned to a cinder. Yet a certain amount of rationality is advisable. It is a distinct advantage when at the dentist's to think of all the good it is going to do to have it out, and not to yield to the very natural supposition that the man behind the forceps is a murderer bent upon doing a full day's work. Martyrs have been known to pass painlessly even through death by fire. They had ceased to care what became of their bodies. Back of pain there is always a protest of the will.

This is but another way of saying that all values are moral values, that all philosophy is either moral philosophy or mere chatter. For morality is no cold list of precepts and prohibitions; it is the very core of conscious existence. You may think that morality bores you, but you are very much mistaken. What bores you is the *word* morality, associated as it is with all the evil attempts of the goody-goody to take the joy out of life. Boredom is that distress which is ours when we neither fight nor yield whole-heartedly. But when we say that we are bored to extinction we express a desire rather than a fact. Boredom, in the sense of utter indifference, is never ours. That sort of boredom *would* extinguish us. The kind we actually experience is that dull distress which comes from our inability either to hate or to love, and in its dullness there is a vastness more appalling than anything acute.

The great anguish of the present age is its *disbelief* in its ability either to hate or to love. Our eyes are fixed perhaps too firmly upon our malice and despair. Yet it is not an evil age, as ages go. If it were we could hardly be so dreadfully unhappy about ourselves. The truly evil ages are dead—and I do not believe there ever were such ages. Life was never all beer and skittles, and yet it is always springing up somewhere, if not in our own gardens then in those of our neighbors.

I remember once asking if it hurt watches to open them (there were hunting-case watches in those days) as much as it hurt oysters. An oyster had bitten my finger, and I was not worrying over oyster pains. But watches had never done me any harm.

I was told that watches did not feel, and consequently did not know. So the only reprisals to be looked for, it appeared, were from watch-owners. Not even oysters, it was claimed, were sensitive enough to know a great deal, nor was it generally conceded that mine had bitten me in revenge for my intent to eat it. The idea seemed to be that oysters and their like lived largely in the dark— the region recommended by animalists as a refuge from our own earthly cares. Here action has a short perspective, and the calculation of distant effects is unknown.

Observation tended to give a certain plausibility to this way of looking at things. A great deal that happened appeared to be more or less unintended by the immediate actors. I had n't meant to hurt oysters—not when the fracas started. That was merely an unthought-of by-product of my own appetite. Sometimes I inadvertently trod upon ants. The elephant at my first circus looked quite capable of treating me in the same off-hand but disastrous fashion. So I ceased beating table-legs when they got in my way—but it began to look very dark for God.

Other things were, in part at least, helpless or unintending. They had alibis. If my earthly father made me wince by extracting a splinter from my knee, he could plead that he was only doing it for my good and to keep it from festering. He couldn't help it if I found the remedial operation painful. But God need n't have invented festering. He needn't have let the splinter run in. He needn't have invented pain. If He was good—and I knew that something was good, for it surrounded me most of the time—there must be Somebody Else, who was evil. Thus I discovered the devil.

He was God's alibi—rather a comfort in his way. And whether you think that there is a devil or not, you must admit that things happen in this world rather as if there were a big one. Even to-day, when we are so often being told that there is neither good nor evil, but only reactions, he seems to be somewhat in evidence. Not a few of these reactions are hellish.

But as a child I knew that wills could be thwarted. There was my own. Being thwarted did not necessarily mean yielding, nor even having one's ultimate purposes annulled in the end. It was my body which sometimes had to yield, as when some other body picked it up, and— no matter how inopportune the moment—put it to bed. Yet, though I was compelled to go along, I did not have to become a party to the outrage by giving my consent. I was free to will, even if I had to take it out in willing and could get nothing and nobody to listen. If that was God's situation, He might continue to be good, even though the thunder rolled and scared me, or the lightning struck the house. As a matter of fact, lightning never did strike the house—not our house—and was amazingly beautiful to look at. Take it all in all, this was an incredibly lovely world. Its dark spots, however, could not be denied, and these seemed to put an end to the idea that God was all-powerful as well as all-good. The devil was not only an alibi, but a rival.

"But God *created* the devil," it was whispered.

"Then He is responsible," was the answer.

"But it was a good devil at first, a shining archangel called Lucifer. God gave him a free will."

"That's where God made his mistake."

"No, it was Lucifer's mistake. He willed to go wrong."

"He ought to have been given a will that couldn't go wrong."

"Then it wouldn't have been free."

"No matter."

"But then it wouldn't have been a will. And if God hadn't created wills, you would n't have had one. You wouldn't have had any conscious existence, for you would not have been able to differ from anything else."

This made creating a very interesting and ticklish business, which began by permitting something to be not yourself. If you did that, it might turn against you. Also it brought the devil home—this idea that he was made devilish by his own will. I could become a little devil myself. Neither the outside nor the inside was all of one mind. Purpose seemed to work itself out only in general averages beyond my calculation. God had not taken me altogether into His confidence, and had certainly neglected to consult me. And evil seemed to result from the operation of other wills which He had turned loose—a tremendously daring proceeding, but the only one which could have made life real.

Fortunately, I wasn't altogether responsible for the consequences, though I have since had friends who considered themselves responsible to a simply alarming degree. My blame couldn't be measured even by the frequent apparent completeness of my own downfall. Evil spirits had a way with them. All I needed to worry about was my own, and usually very tiny, contribution.

On the other hand, those who argued, "It was impossible to exert more will, there wasn't any more will to exert," were obviously wrong. They denied the very nature of Will, which is its mysterious wilfulness, its creative power, its utter omnipotence in the matter of giving or withholding its consent. They deprived it, theoretically, of its power of choice, the only thing of which it consists, though I have never met even a philosopher who was mad enough to do this in practice. So I have begun to wonder, what is the sense or dignity of a philosophy which cannot be put in practice? Or rather, isn't that a man's real philosophy which he does put in practice, and the rest mere humbug? Let those who do not believe in the will stop trying to assert themselves.

For that is what will is, self-assertion. It might be described as the self in action.

"Memory, mind, and will belong not to themselves but to me," said Saint Augustine. "It is I who remember by memory, understand by mind, love by will."

What an odd, old-fashioned notion—that we can Will to Love. One hears oftener of the Will to Power. And yet, if we can will to rule we must be able to will to serve. Such words have a smug sound to many ears, just as real pearls are beginning to have a cheap, artificial look to many eyes in this great Age of Counterfeits. The world has too long been glutted with counterfeit piety. Too long have human ideas been offered as genuine gods. Yet God is not a human concept, though of course my idea of Him is.

As to service, there is no escape from that. Serve we must. It is only a choice of masters. If we fall out with Paul we do but kneel to Apollyon.

But this, you will say, is a child's world which I have been painting. So it is. The time has come to put away childish things. Let us begin, then, to grow up. And for a start we can do no better (or if you prefer, worse) than to turn our attention for a bit to the world of things Jimsian, as described in the modern but well-thumbed Gospel by William James, the unmathematical but unmistakable precursor of Einstein—James the but-half-acknowledged father of the new New World—James, whose fame leaves the greater part of him still unsung— James, the super-journalist of our times—James, in whom so many of us all unwittingly live, move, and try to have our mental being.

CHAPTER II

THINGS JIMSIAN

I. PRIMARY STUFF

IN THE thought of the day, all roads lead to Einstein. Few of us, however, realize how completely Einstein lay, already implicit, in William James. Apart from its proper and mathematical uses, the theory of relativity is merely an attempt to abolish standards. It condemns them all as "arbitrary," and seeks to deprive even thought itself of its very foundation. Theoretically, there are no longer any principles save the principle that there is no principle. Practically, the individual is plunged into that excessive liberty which, though it might spell freedom on a desert island, engenders nothing but slavery when practiced in crowded neighborhoods—the tyranny of spying neighbors, each one an unchained ego in his own right. The world of to-day. But this is precisely the world justified by the philosophy—or rather the several philosophies —of William James.

If it be true that all men seek to kill the thing they love, then love may account for the attitude of James towards the Absolute. It haunted him all his life like an evil dream.

"Let the absolute bury its absolute!" he declared in one of his Oxford lectures.

"Damn the absolute!" he unburdened himself more feelingly to his friend and colleague, Professor Royce, one bright morning at Chocorua.

Now whether one ought to damn the absolute or not depends upon what one means by the word. If you mean your own idea of the incomprehensible, damn it by all means. But if you are peeved by the fact that there *is* an Incomprehensible, your damn is apt to fall back upon your own head—with disastrous results to the noodle. This is exactly what happened to James.

It is now some quarter of a century since he first published his famous essay, entitled, "Does Consciousness Exist?" In it he assured us that there is "no inner duplicity in the primary stuff of the world." No inner duplicity —that is, no real difference between one thing and another. So naturally he found that consciousness does *not* exist—not in any sense which distinguishes it from unconsciousness. Here is a perfect example of that curious insight immediately superseded by total blindness which was so characteristic of James.

He felt drawn towards God. There was enough of Emerson in him to make him feel the pull of some strange Unity at the root of all things. But he was so keenly alive to the diversity of the world, to the individuality of the people he met, to the variety of incident (so appealing to his highly developed newspaper sense) observable in and about Boston and New York, that his very soul revolted at the Emerson-Brahma suggestion of unreality thrown about these differences whose clash compose the drama of actual life.

So he begins by exalting Unity—and how cleverly he does it! There is no *duplicity* in the primary stuff. The act of creation, separating the creature from the Creator, becomes at once an act of treachery, morally bad. Therefore, there never was an act of creation, no *Fiat Lux*. The Great Stuff remains primary, undiversified. No wonder he denied consciousness. For even the Stuff to become conscious of itself would have to split in two. There would then be subject and object—duplicity with a vengeance.

James seems to have been quite unaware that in denying its duplicity he was exalting his Stuff into that very Absolute which he damned. Far worse, he was making the stuff up out of his own mind. So the Absolute became not the Beyond-James, but James. He presumed to say things about it.

And having exalted this Deity who was no Deity, he must needs destroy it in the next breath. There was no inner duplicity in the primary stuff, and so there was no such thing as consciousness—as indeed there wouldn't be under the circumstances. But all the time James was fairly reveling in consciousness. What was to be done about it? The best thing seemed to be to say that consciousness does not exist as an "entity," but to let it go on existing as a "function."

It wasn't an "entity," no—not the sole entity, at least. That wouldn't have left it anything to be conscious *of*. But of what could it be a "function"? Of the primary stuff, of course, since there was nothing else to function. But the function of anything is what it does. The function of a tail is to wag; that of a pump to draw water. Therefore when a primary stuff begins to function, it is no longer primary. We have action; actor, and thing acted upon; cause and effect. No flat contradiction in terms ever landed a philosopher in a worse nest of duplicity than this.

But James was ever bound to run to one extreme or another. He knew of no alternative except alternating opposites. He had to deny unity, or he had to deny diversity. He was all, or he was nothing. To him balance meant compromise—a vicious thing. The All was knowable, or else all was unknowable. But let us see what becomes of a pump's function in a world wherein there is nothing but pump.

We can't of course think about such a pump, for there will be no we, nor any thoughts save in so far as they also are functions of pump. The pump itself will have no shape or limits. It can't be hollow, nor have any inside or outside, let alone the ability to draw water. There won't be any water. All will be pump, an all-pervading, all-including pump which does not pump and is not a pump since it does not differ from not-pump, there being ..o not-pump for it to differ from. Or if you prefer to drop the pump, then there will be nothing but the function, pumping; and as there obviously won't be any pumping, it follows that there will be nothing—which is precisely what the pump was.

But some will say it was unfair to take a pump. I should have taken the primary stuff whose function is not pumping but consciousness. Very well. Here is the primary stuff. Now remember. There is to be no duplicity, no doubling, no twoness. All is to remain simple, and primary, and one. Let us watch the stuff function. What happens? Nothing happens. We ourselves are the primary stuff, and nothing else, so we don't watch. The primary stuff doesn't watch. The primary stuff doesn't function, since it already is its function, consciousness, to wit. Therefore, the primary stuff is consciousness. But this consciousness is not an "entity," an entity being, my dictionary tells me, "anything that exists, or is supposed to exist." So this primary-stuff consciousness is a non-existent consciousness. What James offers us is simply stuff and nonsense. Yet his contemporaries—many of them—believed that he was offering them God.

Even to-day, if you offer the Jimsian dogma in its original vestments to a certain type of audience, there will be a comfortable settling back into seats which have suddenly become pews, and you will hear an audible purring, as over a pious morsel. He had the happy faculty of charming the unwary Fundamentalist while at the same time leading the Unbelievers' parade. The trick of weaving godly words into ungodly sentences.

But where, exactly, did he go wrong? To say that the primary stuff functions as consciousness, isn't that the same as saying that there is an all-conscious Deity? The early fathers of the Church used to define God as *Actus Purus,* or Pure Action. Isn't this what James had in mind?

It would need a more reckless writer than I to venture to say what he actually had in mind. He was no great hand at discovering this himself. His emotions—all of which were generous—were forever getting the better of him. And he reflected his immediate environment so instinctively that quite a difference can be detected between his remarks at Oxford, for example, and those which he let fall from American platforms at popular lectures. But I feel certain that echoing the fathers of any church older than New England was the one thing of which he was never guilty.

God as Action is God as we know Him, Deity as revealed in creation, Deity in relation to ourselves. Inner duplicity is already a fact. If you wish to stress the word *pure,* and make it mean something behind action which is not action, an absolute cause of cause, you at once abandon all meaning in the positive sense. Here is a region which the mind may reach *to,* but not *into.* You may love or hate, but you can no longer think. Thought now consists merely in declaring its own inadequacy. Yes, here was where James went wrong. He tried to think the unthinkable, and finding it impossible ended by denying the impossible. As if the world had been created by human thought, and must of necessity fall within its limits!

So he proceeded to thrash from pillar to post, beating his head against one impossibility after another—impossibles which were none of them that divine Impossible, impossible merely to us, but impossibles all of unnecessary and human construction. No wonder he damned the Absolute. His intellect, in straining to encompass it, had already damned itself. It was thus that he was moved to trample underfoot even consciousness, trying to reason it out of existence merely because it is one of those things which man enjoys but could not have brought about and therefore cannot understand.

James's difficulty is so typical, and so many others have come to grief in the same pit, that it might be well to digress a moment and consider the case, say, of Maurice Maeterlinck, who, in his *Life of Reason,* says of God that he is "the Non-Being which is Being par excellence, the Absolute of the Absolute."

The gentle Maurice was probably juggling with words merely for the sake of seeming profound, but if he meant anything he must have meant that God is Non-Being in the sense of being different from any being which we can comprehend—in which case he might better have said that God is *Being* par excellence. He seems to say that God is Nothing. But not even the word Nothing means anything except something which is different from all given things.

Nothing is so impossible that we cannot even mention it without making it something. We cannot affirm anything regarding a void. Non-Existence par excellence would not even support critics. But Maeterlinck is so unmindful of what he is doing that he even cites Saint Denys the Areopagite (whom he comically describes as "at the source of all Christian mysticism") to testify on behalf of the God-is-Non-Being theory. But what Saint Denys said was this:

"The cause of all things is neither soul nor intellect; it has no imagination, opinion, reason, or understanding: it is not reason or understanding, and it is neither spoken nor thought. Neither is it number, order, magnitude, smallness, equality, inequality, similarity or dissimilarity. It neither moves nor is at rest. It is neither essence, nor eternity, nor time. Even intellectual contact does not appertain to it. It is neither knowledge nor truth. Nor is it royalty, nor wisdom, nor one, nor unity, nor divinity, nor goodness, nor even spirit as we know it."

Maeterlinck terms this "agnostic," and so seems to suggest that the Areopagite was in doubt as to whether God was necessary or not. And yet he has just called Him "the cause of all things." Evidently Saint Denys is not in doubt at all. He merely affirms that the nature of deity transcends as well as includes everything "as we know it."

Christian philosophers are kept from falling into the pit of the Absolute by their belief that God is a Person. The word is derived from *persona,* meaning mask. And the interposition of this mask, separating the Creator from his creation, is the very act of creation itself. No sane person thinks to understand; nobody, sane or insane, can really deny it. For we see it, feel it, encounter it every instant. It gives us our own personal existence. After all, the universe is not a mess of sameness. There are differences, since we perceive them. Were our perception the merest fancy, that would be enough. Beyond this, we can describe God only in negatives.

But to speak of God in negative terms is far from affirming that God is a negative Absolute, related to nothing. So when we try to say anything about an unrelated Absolute we are trying to picture ourselves as we would be if we were not; to draw the design of a creation which hasn't been created; to think as we would think if we were unconscious and non-existent. And we impose this task upon ourselves merely because we hate the idea of anything beyond our grasp, and prefer mystification to Mystery.

This was the intellectual predicament of William James. And yet the heart of the man yearned towards mystery, as do the hearts of us all. His reason, with that fatal flaw at the base of its logic, never satisfied him. He was torn in two, achieving an inner duplicity which was at times pathological. At one moment a cold rationalist, so rational and "tough-minded" in his straight projection of imperfect assumptions as to be starkly irrational, at the next he would become a hot-headed (albeit a "tender-minded") fanatic, capable of accepting the most outlandish hocuspocus as gospel. The cause lay partly in the history of that movement which, continuing even to-day, is more his child than it realizes, and partly in the history of the man himself.

2. HENRY'S BROTHER

In his thirteenth year he had a strange attack which, in his *Varieties of Religious Experience* he describes as having happened to "a French correspondent." It is now known to relate to himself, and he sets it forth as follows:

" . . . there fell upon me without any warning, just as if it came out of the darkness, a horrible fear of my own existence. Simultaneously there arose in my mind the image of an epileptic patient whom I had seen in the asylum. . . . That shape am I, I felt potentially. Nothing that I possess can defend me against that fate, if the hour for it should strike for me as it struck for him. . . . After this the universe was changed for me altogether. . . . I have always thought that this experience of mine had a religious bearing. . . . I mean that the fear was so invasive and powerful that if I had not clung to scripture-texts like 'The eternal God is my refuge,' etc., 'Come unto me, all ye that labor and are heavy-laden,' etc., 'I am the resurrection and the life.' etc., I think I should have grown really insane."

Whether this had or had not a "religious bearing," it is not for me to say. Almost anything might have a religious bearing, one would think. But I wish to make no comment as to the experience itself, nor to hazard even a guess as to what it meant to the soul of William James. This is a review of current philosophy, not an experience-meeting, not a history of philosophers, not—oh, above all, not—an attempt to write an advance report of the Last Judgment. The experience as described, however, suggests an attack of nerves, a bad scare, even something unhealthy in the mentality of the philosopher. He nowhere hints that he really *believed* in any of these scriptural quotations, and throughout his life he continued to associate a just realization of man's precarious situation here on earth with disease. When the devil was sick, the devil a monk would be.

"Healthy-mindedness [by which he means irreligious mindedness] is not the whole of life," he declares in the introduction to his work on his father's literary remains; "and the morbid view, as one may by contrast call it, asks for a philosophy very different from that of absolute moralism."

So the religious view, which is essentially the recognition of a power beyond ourselves, is morbid. To be religious is to have epilepsy, or at least the fear of it. The minute you recover your health, and are therefore certain that no trials are in store for you, absolute moralism becomes your medicine. And everybody, sick or well, is to have whatever philosophy he asks for. Asking for it makes it true.

"To suggest personal will and effect to one 'all sicklied o'er' with the sense of weakness, of helpless failure, and of fear," he goes on, "is to suggest the most horrible of things to him. What he craves is to be consoled in his very impotence, to feel that the Powers of the Universe recognize and secure him, all passive and failing as he is."

I can easily see how absolute moralism—by which I understand the doctrine of those who have earned good treatment at the hands of nature and consider themselves in a position to enforce their claims—might be horrible to anybody, unless he was so awfully well as to have lost his senses. We are all infinitely weak in comparison with the powers of the universe, and if there be no Power with which we may make peace if we will, life is a sheer horror and nothing else. James admits that "we are all potentially. . . sick men," that "the sanest and best of us are of one clay with lunatics and prison-inmates." It was jails and asylums that he feared, and perhaps hospitals—

for even absolute moralists have been known to suffer from apppendicitis.

James was never really well enough to admit the existence of himself and of something else. He was afraid that the something else might hurt him. Or, as Julius Bixler puts it,[1] "James felt the need of the kind of religious support which only a monistic view could bring." Thus the question was (still according to Bixler), "Shall we rest on the everlasting arms, or put on the whole armor of God?"

This is a little obscure, but seems to imply that the choice lay between yielding to God and taking on the whole job of being God. Anyway, these were monistic arms, no-duplicity arms, all-is-one-stuff arms that were being sought in this philosophic sick-room.

"There is no doubt whatever that this ultra-monistic way of thinking means a great deal to many minds," James himself confesses.[2] And those who would like to know what ultra-monism is like he refers to "the paragon of all monistic systems—the Vedanta philosophy of Hindustan, and the paragon of Vedantist missionaries . . . the late Swami Vivekananda." And what says the Swami?

"Where is there any more misery for him who sees this Oneness in the universe, this Oneness of life, Oneness of everything? . . . Where is there any more delusion for him? What can delude him? . . . All fear disappears . . . Whom to fear? Can I hurt myself? Can

[1] *Religion in the Philosophy of William James,* by Julius Seelye Bixler, Associate Professor of Biblical Literature in Smith College, Boston; Marshall Jones Co., 1926, pp. 15–16.

[2] *Pragmatism,* New York and London: Longmans, Green and Co., 1909, p. 150.

I kill myself? Can I injure myself? Do you fear yourself?"[3]

To which I answer: Yes, most certainly. Of course I can hurt myself, kill myself, injure myself, above all delude myself. Not to fear myself when I can conjure up that horrible image of the epileptic patient, that fated hour which nothing may be able to prevent, would seem rather foolish.

"I am the One Existence of the universe," the modest paragon goes on. And James immediately responds from the cantoris side of the chancel: "Surely we have here a religion which, emotionally considered, has a high pragmatic value; it imparts a perfect sumptuosity of security."

And in what does this sumptuous security consist? Evidently in the conviction it brings that the everlasting arms in which one is to repose are not there; in the sick man's discovery that he is only hugging himself!

Thus the patient seeks to become his own physician, his own Absolute. But that is unconscious, unrelated, nonexistent. His refuge is a suicide—utter, complete, and alas impossible—the suicide of total extinction.[4]

"Among the philosophic cranks of my acquaintance in the past," said James in his second Hibbert Lecture, delivered at Manchester College.[5] "was a lady all the tenets of whose system I have forgotten except one. . . . The world, she said, is composed of only two elements, the Thick, namely, and the Thin." Vivekananda was evidently of the house of Thin, and James need hardly have warned us "to distinguish the notion of the absolute carefully

[3] Vide, Pragmatism, p. 152 et seq., where Vivekananda is quoted at length.
[4] See Appendix A.
[5] Vide, A Pluralistic Universe (New York: Longmans, Green and Co., 1925), pp. 135–136. The first edition was published in 1909.

from that of another object. . . the 'God' of common
people in their religion. . . the Creater-God of orthodox
Christian theology." [6] The distinction is rather obvious,
even as between the notions.

But I am much intrigued by this lady, with her rampant
inner duplicity. Her "thick" may be interpreted as meaning whatever was in fact created by this God (not notion)
of common people; while her "thin" stands for the creature's individual, supplementary contributions, notions
properly so-called. Emaciation reaches its climax when
a creature fancies these inner inventions as lording it
without, one of them, the thinnest of all, assigned, it
may be, to loll upon the very throne of thrones.

Or "fat" may be understood to mean our sensuous
experiences, and "lean" our subsequent elaboration of
this material in imagination or in thought. Even in this
case the fat comes from without, and warns us not to
starve ourselves. I might even go on to suggest that religious experiences themselves can only be fattened as they
are fed through the spiritual sense. But most of us prefer
to feed cannibalistically, upon those notions of our notions which we sometimes call Philosophy, or upon those
thrills which we call our Art.

Of these two, our art is certainly the more nourish-
ing. I don't mean that we are very strong in producing
it. The average citizen produces less and less. But we
patronize it—especially music, which we can take with
our meals, canned or otherwise, while the musician does
the work. Art has this advantage over both philosophy
and religion—it demands almost no effort on the part of
the patron. It lets us live in the aspirations of nobler
men, without the bother of trying to put them into prac-

[6] *Ibid.*, p. 111.

tice. And if art degenerates under such circumstances, still it is something. The worst sort of music remains at least a noise.

James is always roughly lyrical, his philosophy but an inner cry. "The history of philosophy," he said, [7] "is to a great extent that of a certain clash of human tempera-ments. . . . His [the philosopher's] temperament really gives him a stronger bias than any of his mere strictly objective premises. . . . The one thing that has *counted* so far in philosophy is that a man should *see* things, see them straight in his own peculiar way."

Truly a strange way of seeing straight, to see so peculiarly as this, though of course it is only our own seeing, straight or crooked, which counts for us in the end. And yet there was once a philosophy built along more scientific lines, each observer contributing his mite, each discovery checked and compared with other dis-coveries, each experience with other experiences, until there was a great body of knowledge—not complete, not ever to be completed, but fairly consistent as far as it went and vastly wise beyond the wisdom of any untutored individual. It was an art, too, since each contribution was due to the inspired light of some single soul. But the modern thinker, as James here not only confesses but boasts, though professing to worship Science, has become sadly unscientific. He sees things—but so does the de-lirium tremens patient. For myself, when it comes to thought, I vote for science, first, last, and all the time. I should even like to see a little more of it in Science.

But since James inclines towards art, stressing the personal equation even until it becomes inartistic, it is necessary to dip into his biography. The musician mani-

[7] *Pragmatism*, pp. 4–8.

fests a part of his individuality in the way he composes a
symphony, but he may be yet more original in the way
he eats his soup. As Boswell says, quoting from Plu-
tarch's life of Alexander, "very often an action of small
note, a short saying, or a jest, shall distinguish a person's
real character more than the greatest sieges, or the most
important battles." Since it is bias and not sense that
counts, it ceases to be of moment what James thought.
We want to know how he "got that way." The great
question immediately becomes, "Did he keep a dog?" and
things like that.

I am inclined to say that he did not keep a dog, not as I
"see things." For he tells us: [8]

"A baby's rattle drops out of his hand, but the baby
looks not for it. It has 'gone out' for him, as a candle-
flame goes out. . . . It is the same with dogs. Out of
sight, out of mind, with them."

You see, the rattle goes out of the baby's sight and out
of the baby's mind at the same time. No baby was ever
known to cry for a missing rattle. As for a dog, when
his master has gone "out" he has gone out, and that is
all there is to it. Who ever heard of a dog pining for
an absent master?

James was forever saying that he was "uncomfortable
away from facts," that people of his kidney always "cling
to facts." [9] These are the kind of facts he usually had in
mind. Here is another sample, this one culled from Santayana, whom James called in as a witness in the case of
the dog.

"If a dog," says Santayana,[10] "while sniffing about
contentedly, sees his master arriving after a long absence,

[8] *Pragmatism,* Lecture V, pp. 174–175.
[9] *Ibid.,* pp. 67–68.
[10] *The Life of Reason: Reason and Common Sense* (1905), p. 59.

the poor brute asks for no reason why his master went, why he has come again. . . . All that is an utter mystery, utterly unconsidered. Such experience has variety, scenery, and a certain vital rhythm; its story might be told in dithyrambic verse."

Santayana's verse is decidedly more dithyrambic than W. J.'s, but evidently it deals with the same dog, one that—though his master is "out"—sniffs about "contentedly"—at the scenery of experience, no doubt, and with a certain vital rhythm. He has no time to consider mysteries. Obviously a pragmatic dog, a philosopher's dog, not a bit like your dog, whose eyes, if you start to leave him behind, ask questions fit to break your heart. Were it not for the sniffing, I should say that the James-Santayana pup was of porcelain.

Anyway, these dog stories have a moral. They show that philosophers' eyes are made of stone—philosopher's stones, if you will, but blind to the ordinary facts of life. This accounts for some of the theories they spin. And in their hands philosophy becomes an idle dream, its art the art of fiction of a highly romantic variety.

Santayana, who knew James well when both were at Harvard, describes him as one who "kept his mind and heart wide open to all that might seem, to polite minds, odd, personal, or visionary in religion and philosophy." "He gave," Santayana continues, "a sincerely respectful hearing to sentimentalists, mystics, spiritualists, wizards, cranks, quacks, and impostors. . . . The lame, the halt, the blind, and those speaking with tongues could come to him with the certainty of finding sympathy. . . . Thus William James became the friend and helper of those groping, nervous, half-educated, spiritually disinherited, passionately hungry individuals of which America is full." [11] He "played havoc" with the earlier "genteel tradition," founded upon Calvinism, and "had a prophetic sympathy with the dawning sentiments of the age, with the moods of the dumb majority." He was "democratic, concrete, and credulous."

And what had he to offer these passionately hungry individuals, of which America and all the world was and is so full? Simply their own spiritual disinheritance. No shepherd, he was a very representative sheep, who could sincerely discourse upon the beliefs of our physical and spiritual ancestors as follows:

"God and his creatures are *toto genere* distinct in the scholastic theology; they have absolutely nothing in common; nay, it degrades God to attribute to him any generic nature whatever; he can be classed with nothing."

Evidently James was at the moment on bad terms with the Jimsian Absolute, so he saddles it upon Saint Augustine, Albertus Magnus, Saint Thomas, and Dante. The Creator-God didn't belong to these people, they were too uncommon. And the common people themselves are beginning to be no longer "distinct" even from that God which is theirs. They run together, like gobs of half-baked dough. I fear that the "mystics" with whom James consorted were half-baked mystics. Nor is it to be inferred that the scholastic Supreme Being is here classed with "nothing" in any savingly subtile sense. He is simply "out."

"There is a sense, then, in which philosophic theism makes us outsiders and keeps us foreigners in relation to God, in which, at any rate, his connection with us ap-

[11] *The Winds of Doctrine,* Ch. VI.

pears as unilateral and not reciprocal. His actions can affect us, but he can never be affected by our reaction"—all of which would seem to indicate that the earlier Christians refused to believe that man was created in God's image, or that there was any joy among the angels over even a dozen sinners who repented.

"God is not the heart of our heart and the reason of our reason, but our magistrate rather,"—and if Thou, Lord, shalt be extreme to mark what is done amiss, O Lord, who may abide it? "In scholastic theism [the Scholastics have anyway escaped now from being "philosophic" theists!] we find truth already instituted and established without our help, complete apart from our knowing; and the most we can do is to acknowledge it passively and adhere to it,"—and it is going to be some job to adhere to that particular part which is apart from our knowing.

"In scholastic theism . . . truth exists *per se* and absolutely . . . no matter who of us knows it or is ignorant, and it would continue to exist unaltered, even though we finite knowers were all annihilated." Which is to say that the scholastic theists taught that the creation made no alteration in things, that all would have been just the same without any creation!

"This dualism"—for the Absolute has suddenly become dual—"and lack of intimacy has always operated as a drag and handicap on Christian thought." [12] It must have been. So James substitutes the intimacy of his own everlasting arms. And this lecture, this burlesque of some two thousand years of history—of our *own* history—was not written by a Hottentot but by a professor at Harvard, delivered to the student body at Oxford, then printed in a

[12] *Vide, A Pluralistic Universe,* pp. 26-29.

book and translated into all modern languages. And no
one yet has been moved to tears. How account for that?

Plato says in his *Critias* that "upon an audience of human beings it is easier to produce the impression of adequate treatment in speaking of gods than in discoursing
of mortals like ourselves. The combination of unfamiliarity and sheer ignorance [in such an audience] makes the
task of one who is to treat a subject towards which they
are in this state easy in the extreme. . . . Where the subjects [of discourse] are celestial and divine, we are satisfied by merely faint verisimilitudes; where they are mortal and human, we are exacting critics."

Not so very exacting, either, when the verisimilitudes
of philosophic dog-and-baby stories go down without a
ripple. What wonder, then, that the makers of discourses
upon subjects celestial and divine hesitate at no monstrosi-
ties of unfamiliarity and sheer ignorance whatsoever?
And come to think of it, this discourse of James was not
of gods, but merely of saints and doctors.

To make the case more incredible still, James was born
—not in Timbuktu, but in New York, and in 1842, when
the Bible Belt still extended in America to the Atlantic
seaboard. His father, Henry—William was to have a
brother Henry a year later, and a son Henry in due time—
was a noted theologian. Apart from the conversation and
example of a highly intellectual home circle and the assistance of private tutors, William's educational opportunities included such trifles as the following:

One year at the University of Geneva; one year as art-
student under Hunt at Newport; two years at the Law-
rence Scientific School; two years at the Harvard Medical
School; one year of scientific field-work under Agassiz

in Brazil; one year in Berlin, studying psychology; and finally, graduation from Harvard with the degree of M.D., in 1876.

Harvard made him Assistant Professor of Philosophy in 1880, and full professor five years later. His teaching career at Harvard lasted no less than thirty-five years, the first seven devoted to physiology and anatomy, the next nine to philosophy, the next nine to psychology, and then ten more to philosophy. He lectured at Columbia and at Oxford, and when he died, in 1910, the list of his honors filled a column in *Who's Who*.

Not exactly an ignorant man, one would think. And yet —did you notice anything peculiar about his training? He was a doctor of medicine, who became a psychologist by dint of being an anatomist and physiologist, a philosopher whose philosophy was a graduate course in psychology. He learned his business from the ground up. But philosophy is a business that ought to be learned from the sky down. Why? Because effects proceed from cause, and not causes from effect. Philosophy is above all things the study of Cause. It must meet the test of effect—or as we say, of facts—but he whose twig is bent by the weight of matter seldom becomes a tree of the higher knowledge.

But there is no use in preaching this doctrine, for, as G. K. Chesterton somewhere says, the most conspicuous characteristic of the present age is its downside uppishness, or as he puts it, its tendency not merely to let the tail wag the dog, or to dock the dog of its tail, but to to dock the tail of its dog. As usual, Chesterton slanders us. Why, even in the days of William James we had already learned to dock the tail of its tail, and then to construct our philosophy out of the remaining wag. Thus, though the physicist cannot so much as pour sulphuric

acid out of a bottle without having been first created and
entered into some sort of relation with his Maker, cannot
so much as observe either acid or bottle without assuming
a multitude of conclusions as to the nature of knowledge
and perception, which are problems of epistemology entirely this side of physics, we continually try to found our
thought upon atoms.

Not even the sequence of his studies, however, can fully
account for James. It was rather his intellectual disinheritance which moulded him. Brought into contact as he was
with what looked like the culture of the ages, he remained
to the end a philosophical backwoodsman. Culture of this
sort simply would not "take." The influence of his pioneer
countrymen rose about him like a screen. His heart was
in the wilderness there with Esau. This is the secret of
his immense significance. He had "a prophetic sympathy
with . . . the moods of the majority." Nor did he revolt
against the older world of thought so much as against that
thin "genteel tradition" which those sentimentalists, wizards, cranks, and quacks had been taught to regard as the
older world of thought. His God had been their God, and
so now his godlessness was to become their godlessness.
The agony of his inner discord made him an artist after
his kind, intimate, personal, and yet representative. And
as he escaped that Anglicizing which so often turns the
American pioneer into a mere provincial, he is respected
abroad, like Emerson, Whitman, Poe, Edison, and Henry
Ford.

3. THE CAT'S PURE OTHER

James devoted his whole life to an attempt to discover
something "thick," and real. He was like a man who,

looking at one side of a slice of bread and finding it a mere surface, turns to the other side. But both surfaces are appallingly thin, and none shall have bread who ignores the one while he clings to the other. James repudiated the "intellectualism" that says that bread is all topside, but his "empiricism," or totally bottom-side baking, was just as unsubstantial. In spite of the many splendid qualities of his heart and mind he went hungry. Not even his famous loaf of Pragmatism had any logical substance.

The word was first introduced into philosophy by Mr. Charles Pierce, in an article entitled, "How to Make Our Ideas Clear," published in the *Popular Science Monthly* for January, 1878, and comes straight from the Greek *pragma,* meaning action, from which the words "practice" and "practical" are derived. As a slogan-maker, Pierce was certainly in the upper brackets. Give a dog a good name, and who dares to kick him? You would not be impractical, would you? Therefore, so it seems to a certain type of mind, you *must* be a Pragmatist.

Nor was Pierce's fundamental idea—that there is no such thing as a difference which does not make a difference to something or somebody—anything less than an inspiration. If he had only succeeded in making his own idea clear, so clear that no one else could have muddled it! But in this he failed. And soon after James took it up it began to be rumored that Pragmatism really means: If you don't *think* there is a difference, there isn't any. Things are whatever you *think* they are. Your thinking—and more especially mine—makes them so. And I think whatever I like. It follows that you believe that a true thing's truth comes from your thinking it true, so you really believe that it is not true. Therefore Pragmatism itself, according to Pragmatism, is untrue if you believe in it. You see how "practical" it is.

Santayana, in his *Winds of Doctrine,* says that one "good critic" has discovered "thirteen pragmatisms." If he had been industrious as well as good he might have discovered thirteen hundred. There is no limit to the number of things that you can think. In this sense, all philosophies are pragmatic, even those which are not. Nor will they all be foolish. To oppose everything which is uttered in Pragmatism's name would be like opposing vegetation. Not every plant is poisonous. So the question is, of what sort was the Pragmatism of William James?

He of course resented all such definitions as the one I have given, which has the defect of being intelligible, and the further defect of being just. As early as 1906 he branded as "impudent" all those "persons who think that by saying whatever you find it pleasant to say and calling, it truth you fulfill every pragmatic requirement." They were guilty, he insisted, of "slander" in addition to impudence. If so it must have been because of the ancient rule of law which holds that "the greater the truth the greater the slander," for in a lecture entitled "What Pragmatism Means," James himself declares:

"Day follows day, and its contents are simply added. The new contents are not true, they simply come and are. Truth is what we say about them." [1]

It isn't, then, even necessary to think. Talk will do. To-day's contents included a thunder-shower in the early morning, but that simply came and was. I had the presence of mind to remark, "It's a fine day." And lo, there wasn't

[1] *Pragmatism,* p. 62

a cloud in the sky! Not only that, but there hadn't been a cloud in the sky.

"The finally victorious way of looking at things will be the most completely impressive way to the normal run of minds." [2]

If I am out-voted I shall cease to be normal, and if the majority are not sufficiently impressed with sunshine it may rain yet.

"The true, to put it very briefly, is only the expedient in the way of our thinking, just as the right is only the expedient in the way of our behaving." [3]

So the Right was never nailed to a cross. Give unto us Barabbas, the Normal, the Expedient! It would be difficult, I think, to slander this philosophy. Moreover, the earth was once victoriously and impressively flat, and impudent, abnormal minds like Galileo and Columbus were wrong in thinking otherwise. Then the normals took to touring until finally they decided that it would be more expedient to think in terms of circumnavigation, whereupon the earth got around to being round—whether permanently or not time alone can tell.

However, there is hope for minorities. You have but to say that other people are abnormal; and then, if you like, you can make it square, and become your own victorious majority.

"The world is indubitably one if you look at it in one way, but as indubitably it is many, if you look at it in another. It is both one and many—let us adopt a sort of pluralistic monism." [4] A sort of philosophical round-square.

You think he merely intended to suggest that the world

[2] *Ibid.,* p. 38.
[3] *Ibid.,* p. 222.
[4] *Ibid.,* p. 13.

is one to *me* and as far as I am concerned, if I look at it that way; and as indubitably many to *me* and as far as I am concerned, if I look at it in another? How very charitable you are. But he had not the air of meaning this, nor would even this be true unless I could keep on thinking that I was always right. And this would be difficult. What if I think that arsenic is sugar? There comes a bitter taste, palpitation of the heart, gastro-enteritis, and possibly cerebro-spinal symptoms. Comes, in fact, the ambulance, and perhaps the morgue. Of course, if I continue victoriously to think that the arsenic was sugar, it remains sugar so far as I am concerned. Nevertheless, it will require considerable effort to believe that I am still sitting at my tea with nothing in particular going on. Is there not, then, *some* difference, after all?

"Truth," says James, "is a property of certain of our ideas. It means their 'agreement,' as falsity means their disagreement, with 'reality.' Pragmatists and intellectualists both accept this definition. . . . [They] begin to quarrel only after the question is raised as to what may precisely be meant by the term 'agreement,' and by the term 'reality,' when reality is taken as something for our ideas to agree with." [5]

It must be remembered that "Intellectualist" and "Idealist" were merely swear-words with James, and signify non-Jimsians. But he won't quarrel with what they say until it is understood to mean something, in which event he must challenge the term "reality" because it suggests an outer reality, a something not James. Pragmatism shrinks, he says, "from looking backwards upon principles, upon an *erkenntnisstheoretische Ich,* a God, a *Kausalitätsprinzip* . . . taken . . . as something august

[5] *Ibid.,* p. 198.

and exalted above facts. . . . Pragmatism shifts the emphasis, and looks forward into facts themselves." [6] This enlarges the field of objection so as to include not only any God who is not James, but any James who is not God—that is, any James who is due to a cause above the mere fact of there being a James.

And what is his quarrel with the term "agreement"?

"The popular notion is that a true idea must copy its reality. Like other popular views, this one follows the analogy of the most usual experience." [7]

Well, it has not been my most usual experience—not since I began to devote my time to these unrealistic philosophers. And one must lament here a touch of demagoguery, that old trick of flattering the majority by addressing them as if they were the select and unappreciated few. But he admits that "our true ideas of sensible things do indeed copy them. Shut your eyes and think of yonder clock on the wall," he says, "and you get just such a true picture or copy of its dial."

But this picture is not an idea, but a memory-image. So James plunges into his real quarrel by alleging that "when you speak of the 'time-keeping function' of the clock, or of the spring's 'elasticity,' it is hard to see exactly what your ideas can copy. . . . Where our ideas cannot copy definitely their object, what does agreement with that object mean? Some idealists . . . speak as if our ideas possessed truth just in proportion as they approach to being copies of the Absolute's eternal way of thinking. . . . The great assumption of the intellectualists is that truth means essentially an inert static relation. When you've got your true idea of anything, there's an

6 *Ibid.*, p. 122.
7 *Ibid.*, p. 199.

end of the matter [say the intellectualists]. You're in possession; you *know;* you have fulfilled your thinking destiny." [8]

The "intellectualist" is here represented as a fellow who
believes that if I think it is three o'clock, and it is in fact
three o'clock, I am *through,* and must keep right on thinking three even when it is after four. The enemy's clock
is identified as that famous grandfather's clock which
stopped short, never to go again, when the old man died—
a dead clock, whose spring's elasticity and time-keeping
function are the ideas of an Absolutely Dead God, fit
only for a very dead grandpa. To preserve a static relation
with such a piece of machinery, is certainly to be through.
James was very wise to turn his back upon an intellectualism such as this. His slight mistake was in supposing
that the whole world was sunk in a petrified theology,
essentially Calvinistic, that he and his followers alone had
escaped.

But to speak of the Absolute's "eternal" way of thinking, or of an Absolute as "thinking" at all, is really a too
great abuse of words. So what if we substitute a living
Infinite, and a clock that goes? To preserve anything like
a static relation of agreement with *this* time-piece, our
ideas too will have to be on the move. Or can an idea keep
in touch with nothing but paralysis? James is fighting a
straw man whom he has hurled into the old James-made
void. And having discovered that such an Absolute can
neither move nor think, he has great sport in making Mr.
Straw's "reality" depend upon this impossible thought
and motion. But as we are now dealing with the world of
human experience, we ought in common fairness both to
ourselves and others to drop this Absolutely Inane. We

[8] *Ibid.,* pp. 199–200.

are not immediately concerned even with that well of Being of which we can allege nothing but metaphors. We are in the world of time, of clocks. Creation has, somehow, come about.

In his own keen awareness of this, in his realization of the continual creativeness of creation, lies the chief gift of James to the world. And he was so bucked up to find Henri Bergson plucking at the same string that he gave the Frenchman altogether too much credit as the originator of the theme. Had his geographical sympathies been a little broader, had his historical sense reached a little further back, he might have discovered what had been discovered long before and not everywhere forgotten— that this creativeness is not really *in* creation, but in the Creator; and that creativeness is in its very essence miraculous, beyond all thought, and therefore Absolute in the sense of being unrelated to any possible human idea of it—unless we regard mere otherness as a relationship.

Even without becoming a veritable sage, James might have found something for our ideas to copy when they cannot copy their subject. Were things static, frozen, in a changeless present like tadpoles in a cake of ice, an idea would be hard put to it. But ideas are concepts, workings over of remembered perceptions. They strive to copy the future. When I say to myself, "Yonder clock is keeping time," I conceive of motion, something common extracted from my experience with a great many moving things, but not necessarily accompanied by any mental picture of a particular moving thing. I am by no means through when I think, "Now it is three o'clock." I go on and say, "It must be three-ten." Then I open my eyes and compare what has now become a mental image, with a visual impression. But first my idea of motion in general has come from comparing one remembered sensation with another. Were there no opportunity for ever comparing anything with anything else, I could never entertain the idea of motion, or have any ideas whatever.

But see how James, whose chief idea seems to be that Deity, if any, would necessarily be a solid lump, tangles himself up in this simple situation. "The truth of an idea," he trumpets forth, "is not a stagnant property inherent in it," though this is just what it would be if it were created by my say-so, and I should happen to start repeating myself.

"Truth," he continues, *"happens* to an idea. It *becomes* true, it is *made* true by events. Its verity *is* in fact an event, a process : the process namely of its verifying itself." It verifies *itself,* you see. Delightful task! to be a tender thought, and teach your confirmation how to shoot —as Thompson might have said.

But look again at yonder clock. I am still thinking it is three, but the hands in truth point to four-thirty. Truth has not yet happened to my idea, it has only happened to the hands. But it will happen to my idea if I am patient. I have only to stick to it until three o'clock to-morrow morning. But it looks to me as if even then my idea will be verified by the hands, not the hands by my idea.

Or let us suppose that you are my lawyer, and that I have a business deal which will take us both to New York City. You live in Port Jervis, and agree to take the Orange County Express on the Erie on a certain morning. I live, let us say, in Middletown, N. Y. (I did once), and you are to pick me up there as you go through. Your watch is slow, but you have a belief that it is right.

Now, if I were a Pragmatist, I should not know what that statement "means," and we should begin to quarrel

over the signification of the term "right," and the term
"term." As things stand, I know very well without quar-
reling. It means that you are going to lose your train.
That, however, makes no difference to me. You haven't
lost it yet. Nor does your mistaken notion about your
watch make any difference to me. It isn't even true that
you have such a notion, for I know and say nothing about
it. Truth hasn't happened to it yet. It is one of those things
in the day's contents which simply come and are.

But when you do lose the train, and I am left alone to
handle the deal, and go and lose money on it in consequence, why—what happens then? Is it now true that
your watch is slow? True that you thought it was right?
Apparently. But truth hasn't exactly happened to the idea
that you did entertain about your watch. It would almost
seem as if error had happened to that. Stranger still,
truth has happened to the notion that you were mistaken
about the reliability of your time-piece—and that was
nobody's notion. Nobody entertained it. Neither you nor
I dreamed of such a thing. So along comes truth and happens to an idea which did not exist. That is the *erkenntnisstheoretische, Kausalitätsprinzip* fact of the matter.

"But what do the words verification and validation
themselves pragmatically mean," now that they have happened to the never was? "They . . . signify certain practical consequences of the verified and validated idea. It
is hard to find any one phrase that characterizes these
consequences better than the ordinary agreement formula
—just such consequences being what you have in mind
whenever you say that our ideas 'agree' with reality. They
lead us, namely, through the acts and other ideas which
they instigate, into, or up to, or towards, other parts of

experience with which we feel all the while—such feeling being among our potentialities—that the original ideas remain in agreement. The connections and transitions come to us from point to point as being progressive, harmonious, satisfactory. This function of agreeable leading is what we mean by an idea's verification." [9]

James at least has found something of our ideas to agree with! But he would have it that we are led by this agreement. We hold an idea, and the consequences to be deduced from it prove to be precisely what actually happens. Ergo, the idea was true. True enough. But as this agreement, the instant it happens, belongs to the past, it must be, according to James, that we are led by what is behind us. Why didn't he say that it was hope or belief that led us? Because he was trying to prove that the idea wasn't true until it *was* verified, while at the same time holding that it was true, whether true or not, if we thought it true. He was endeavoring to convince us that truth and our knowledge of the truth are one and the same thing.

At the same time he is leading into, or up to, or towards the doctrine that agreement is necessarily agreeable. So it is when it is agreement with our will. But if I had an idea that I was about to be lynched, and it turned out that the crowd around my door were merely bent upon singing me a serenade, the non-verification of my original impression might be likened to a cloud with a silver lining. Even though "thus wak'd to rage by music's dreadful power," I should seek for no agreement with the idea of hanging. James, however, always has a feeling that truth must be agreeable—not that he assumed that everybody was living in saint-like submission to the Almighty, but because to

[9] *Pragmatism,* p. 202.

believe in disagreeable truth suggests a lack of harmony, and this implies a duplicity in the primary stuff. He clings to Monism, the *sine qua non* of the Is-Not-So.

Josiah Royce, the great pragmatist's tender-minded intimate, was also much concerned with this problem of otherness. And one day he happened to remark that, even for a cat to look at a king, the "cat's idea . . . must transcend the cat's own separate mind and somehow include the king. For were the king utterly outside and independent of the cat, *the cat's pure other,* the beast's mind could touch the king in no wise." I am quoting James's own paraphrase of Royce, as found on page 63 of *A Pluralistic Universe,* to which James adds the following comment:

"This makes the cat much less distinct from the king than we had at first naïvely supposed. There must be some prior continuity between them, which continuity Royce interprets idealistically as meaning a higher mind that owns them both as objects, and owning them can also own any relation, such as the supposed witnessing, that may obtain between them." James found Royce's reasoning "pleasing from its ingenuity." Why, then, not "true"? But no. He merely held it to be "almost a pity that so straight a bridge from abstract logic to concrete fact should not bear our weight," and objected that "the purely verbal character of the operation is undisguised."

I don't claim that Royce's verbalism was altogether happy. I can't quite see a cat's idea including a king, or a king who was not somewhat independent of his cat. Otherwise king would be created by cat. But it is perfectly true that if the king were the cat's pure other, if they had nothing in common, pussy would see only an empty throne. Yet, instead of saying that the cat's idea transcends the cat's mind, why not say that if it were not for something transcending even the cat's idea there would be no idea, no king, and no cat?

As to what the mouser and the monarch had in common, it is quite obvious that it was the material universe. Admit that neither saw quite the same universe, that each had his own private vision. Call it a private illusion, if you will. Yet these private universes of ours are held together by their common origin, so that they largely overlap. These overlaps hang together with a vast consistency, and so cease to be private, or illusory. That is why we speak of the universe as external. Only by believing that I myself invented both stairs and neighbors can I escape from the conviction that my neighbors keep climbing up and down my stairs because they see and feel the steps. And it is into this escape that James immediately plunges, for he says:

"Because the names of finite things and their relations are disjoined, it doesn't follow that the realities named need a *deus ex machina* from on high to conjoin them. The same things disjoined in one respect *appear* as conjoined in another. Naming the disjunction doesn't debar us from also naming the conjunction."

This sounds the way a headache feels, but the suggestion is that no *deus ex machina* from on high is necessary, because a *deus ex machina* from Harvard will answer every purpose. If there had been no context, he might be thought to say that things joined in some respects are disjoined in others, that a cat resembles a king in having four limbs but differs from him in having all of them feet. But this is too uncontradictory ever to be James. He must also play with the shadowy idea that he has but to conjoin cat and king in his mind, or in his vocabulary,

and the cat sees. He has but to disjoin them, and the cat
goes blind. This seems also a little verbal. If two things
are distinct from each other, neither can see the other
distinctly! We distinguish only what is indistinguishable!
I fear we have fallen into the pit of the absolute cat, that
the cat itself is the cat's pure other. And, strangest of all,
it is now that we begin to see a king!

But let us turn to our other pure other, the cow.

"If I am lost in the woods and starved," says James,[10]
"and find what looks [merely looks!] like a cow-path, it
is of the utmost importance that I should think of a human habitation at the other end of it, for if I do so and
follow it, I save myself."

That is, it is of the utmost importance that I should
follow this as-yet-unverified idea whose truth has not yet
happened to it, so that I may be led agreeably, progres-
sively, harmoniously, satisfactorily, from point to point,
and by a verification which is unverified, into, or up to,
or towards, a human habitation. But what if—as seems
very likely, considering the nature of the spoor—the
habitation turns out to be a cow-shed? A cow-shed, of
course, will be sufficiently agreeable if Katy be waiting
for me there in the m'-moon light. But what if it be an
angry bull?

"I have honestly tried to stretch my own imagination,"
James assures us,[11] "and to read the best possible meaning
into the rationalist conception. . . . The notion of a
reality calling on us to 'agree' with it, and that for no
reasons, but simply because its claim is 'unconditional,'
or 'transcendent,' is one that I can make neither head nor
tail of. . . . I try to imagine myself as the sole reality in

[10] *Pragmatism,* p. 203.
[11] *Ibid.,* pp. 134–135.

the world, and then to imagine what I would 'claim' if I were allowed to."

The cat's pure other is out of the bag! He tried to imagine himself as the sole reality in the world. And he discovers that under such circumstances he would "claim" nothing. That is, he would not force his creatures to encounter cow-sheds when they were looking for houses. But, as he is the sole reality, evidently he has neglected to create other realities. There is nothing, then, upon which any claim can be made, neither men, cows, houses nor sheds. It must be that, having put himself in the place of God, he couldn't quite discover the way to go about this business of creating. He concludes that God would be in the same fix, and of no pragmatic value. So he returns to earth, and can make neither head nor tail out of the idea that sheds and houses should not come and go at the behest of his beliefs. He is still the only reality in the universe. And yet, having failed to bring himself about when he was playing God, he has no being.

"Souls," he tough-mindedly informed the students at Oxford,[12] "have worn out both themselves and their welcome, that is the plain truth. Philosophy ought to get the manifolds of experience unified on principles less empty. . . . Like the word 'cause,' the word 'soul' is but a theoretic stop-gap . . . it marks a place and claims it for a future explanation to occupy. . . . You see no deeper into the fact that a hundred sensations get . . . known together by thinking that a 'soul' does the compounding than you see into a man's living eighty years by thinking of him as an octogenarian."

True, the word "soul" does not enable us to explain how a hundred sensations get known together, or at all. It is

[12] *A Pluralistic Universe*, pp. 209-210.

merely a convenient way of recording the fact that they do, of referring to that strange experiece of being alive, of having experiences which we know to be ours. The inmost nature of the soul remains a mystery, just as the inmost nature of "cause" (that is, of God) remains a mystery. In the case of God, His Unknownness is that with which we have not entered into relation, and may truly be described as our Pure Other. In the case of the soul, our lack of knowledge is due to the very opposite circumstance that it is not our other at all, but our self, from which we are not "distinguished." There can be no knowledge when there is no distinction, and none when there is no resemblance. But to mark a place, even a place of miracle with a burning bush, is, it seems to me, a very useful thing to do—unless, of course, you wish to fool yourself with the notion that whatever you do not understand does not exist, in which case it may be more convenient to lose the place and close the book.

James has now refused to make a God of God, and failed to make a God of self. What remains ? The body. We will make a God of Matter. That is, having denied force, we will make a God out of the clash of forces. This is like denying both the upper and the under surfaces of a slice of bread, and then talking about the middle as existing without surfaces. Never mind, it is perfectly pragmatic, and James undertook the job with the assistance of Bergson and Fechner—more especially Fechner.

Gustav Theodor Fechner (1801–1887) was, like James, the son of a clergyman. Like James, he was first educated as a doctor and became a philosopher by dint of ceasing to be a scientist. And like James he in early manhood suffered a terrible nervous experience—amounting in Fechner's case to complete collapse and a painful hyperæsthesis of all the functions.[13] To put it bluntly, he was a little cracked. Yet James devotes the entire fourth lecture in *A Pluralistic Universe* to the praise of his views, and says, "He was in fact a philosopher in the 'great' sense."

The "great sense" of this philosophy consisted in the doctrine that the earth is an animal not inferior but superior to man; that the stars are angels; and that the highest intelligence of all is that of the stellar universe. God is not nebulous, he is merely a nebula.

"Our animal organization," Fechner says, in a book which he presumed to call *Zend-Avesta*, "comes from our inferiority" to the Earth. "What need has she of arms, with nothing to reach for, of eyes or nose when she finds her way through space without either, and has the millions of eyes of all her animals to guide their movements on her surface, and all their noses to smell the flowers that grow? For as we are ourselves a part of the earth, so our organs are her organs."

So James finally got his various experiences compounded and unified without a soul. He let the ground do it. The explanation is no longer "verbal," it is "thick," "tough." Man is a louse inhabiting the fuzzy crust of the planet's great skull. Not even that. He is but one of her organs. She has no kidneys, but she can make use of his. He smells the flowers for the benefit of the noseless hills. A handful of pebbles has a little wisdom, a bushel of pebbles has a little more, and a comet's tail is quite a philosopher. Whatever is bigger, in a physical sense, is better in an intellectual sense. Great is the elephant, great the whale, but think of the awful knowledge possessed by the Milky Way!

Did James really believe this? Of course not. But he

[13] See Appendix B.

thrilled at the sight of these mad landscapes, and for a time almost thought that he believed. He says that he "was envious of Fechner and the other pantheists" because he himself "wanted the same freedom" that he saw them "so unscrupulously enjoying." But his "conscience" held him "prisoner."

"In my heart of hearts, however," he continues,[14] "I knew that my situation was absurd and could only be provisional. That secret of a continuous life which the universe knows by heart and acts on every instant cannot be a contradiction incarnate. If logic says it is . . . so much the worse for logic. . . . I struggled with the problem for years. . . . I saw that I must either forswear that 'psychology without a soul' to which my whole psychological and Kantian education had committed me . . . in a word bring back scholasticism and common sense . . . or finally face the fact that life is logically irrational. . . . For my own part I have finally found myself compelled to give up the logic, fairly, squarely, and irrevocably. . . . Reality, life, experience, concreteness, immediacy, use what word you will, exceeds our logic, overflows and surrounds it."

How almost true all this is! Reality does indeed overflow and surround our mental capacity. But what is there "irrational" in that? Is it contrary to logic that the ocean should overflow and surround the water-holding capacity of a thimble? But James was incapable of "compounding" his own ideas, of letting one idea correct another. His notion of balance was to alternate violently opposite moods and conclusions. His mental health consisted of a

[14] *A Pluralistic Universe*, pp. 207–212.

long series of chills and fevers. So he was wise to give up his "logic," considering what sort of logic it was. But he could not bring back common sense and scholasticism. He could only deliver us over to Bergson.

CHAPTER III

THE WAY OF ALL FLESH

I. BERGSON

SOME will have it that Pragmatism consists in applying the Scriptural test, "By their fruits ye shall know them." How little this is true we have just seen. Nevertheless, it ought to be true. Let us then be pragmatical in this practical sense, and apply the test to Henri Bergson.

He was a man impressed with the genuine newness of the present. This is the supreme merit which shines through his pages, and must never be forgotten for a moment. One of the results of his work, therefore, is to strengthen the cult of change; of the merchant who seeks a quick turnover at a small profit; of the landowner who builds flimsy houses so that it will not cost too much to tear them down; of the woman who prefers a fashionable, imitation fur coat every year to one genuine sable that will last her lifetime.

He casts his vote, then, for the individual as distinguished from society, for emotion rather than thought, for to-day and against yesterday—or, as his disciples might phrase it, for life and against death.

At first and on the face of it, his philosophy even looked like a short-cut back to primitive Christianity, with its belief in the supernatural, a turn of the tide away from the unreality of those metaphysicians who, from being

~ HENRI BERGSON ~
He made irrationality respectable

excessively anti-Catholic had become anti-Protestant in their endeavor to reduce all things to the limits of human understanding.

But the Church would have none of him, and to the plea that here was at least a case where half a loaf was better than no bread, turned a deaf ear. Bigotry? Ingratitude? Mm! Well, no. For the fact of the matter is that Bergson's "life" is spurious, his bread unwholesome, as if salted with a salt from which the sodium has been omitted, leaving a deadly chlorine. It may be swallowed, along with an antidote, without danger to a strong stomach, but that is no excuse for recommending him as a baker.

Real passion breathes through his pages. He quickens the pulse with the ardor of his images. As moving pictures they stir and delight. But they are offered as thought, and to thought and through thought they are fatal. Were it not for this, he might be classed as an apostle of mere drunkenness. A little might do us good, if it were drunkenness of the right sort. But Bergson's quickener of all things, the sensuous *"élan vital,"* destroys the mind that tries to acept it—a very slight improvement upon refusing the body its just sop, in the good old New England manner.

Let his begin, however, by putting his best leg foremost. See with what positive genius he attacks, for example, the intolerable subject of Nothing in the fourth chapter of *Creative Evolution*. One has but to make certain omissions, and read into the text a few things which are not there, and the result is an essay with all the stimulating peace of ancient wisdom. He is speaking of our "theoretical illusions," more particularly our notion that we can think of not anything at all.

"All action," he says, "aims at getting something that we feel the want of. In this very special sense it fills a void, and goes from the empty to the full, from an absence to a presence, from the unreal to the real. We are immersed in realities and cannot pass out of them; only, if the present reality is not the one we are seeking, we speak of the *absence* of this sought-for reality whenever we find the *presence* of another. This is quite legitimate in the sphere of action, but whether we will or no, we keep to this way of speaking and also of thinking, when we speculate on the nature of things independently of the interest they have for us."

Thus we speak of a room as "empty," meaning merely that it is unfurnished, is empty of tables and chairs. We say that nothing happened during the evening, meaning that we were bored, or that the expected, the hoped-for, or the feared did not take place. The "idea of nothing" is but an idea of something other than a particularly specified thing. But if we fancy that we can positively conceive of the absence of all things together, we deceive ourselves.

"The question," for example, "is to know why there is order, and not disorder," chaos instead of the cosmos. "But the question has meaning only if we suppose that disorder, understood as the absence of order, is possible, or imaginable, or conceivable. It is only order" in fact, "which is real; but, as order can take two forms," the kind we are looking for and the kind we are not, "and as the presence of the one may be said to consist in the absence of the other, we speak of disorder whenever we have before us that one of the two [sorts of order] for which we are not looking. The idea of disorder is then entirely practical," in this limited sense. "It corresponds to the disappointment of a certain expectation." But it "does not denote the absence of all order, but only the presence of that order which does not offer us any actual interest. So that when we try to deny order completely, absolutely, we find that we are leaping from one kind of order to the other."

Thus we find the furniture not arranged as we would have it for receiving guests, and we say that the room is in disorder. But of course the tables and chairs are nevertheless arranged in *some* order or other. If we wished to give visitors the idea that there had been a riot, it would be quite in order to have tables and chairs piled in a heap. Disorder is merely an arrangement to which we are indifferent or opposed.

But we are prone to forget how personal and full of bias is our way of looking at things. And so we permit "the problem of knowledge" to become complicated "by the idea that order fills a void, and that its actual presence is superposed on its virtual absence." And in our illusion we say that order in general or the universe comes out of nothing when we are only entitled to say that it comes out of the Not-Anything-Which-We-Can-Understand.

"I have no sooner commenced to philosophize," he continues (I am condensing his text), "than I ask myself why I exist. I want [also] to know why the universe exists. And if I refer the universe to a Principle that supports it or created it, my thought rests on this principle only a few moment, for the same problem recurs, this time in its full breadth and generality: Whence comes it, and how can it be understood that anything exists?

"Now, if I push these questions aside and go straight to what hides behind them, this is what I find: Existence appears to me like a conquest over nought. I say to myself that there might be, that indeed there ought to be, nothing, and I then wonder that there is something.

"Or I represent all reality extended on nothing as on a carpet; at first was nothing, and being has come by super-addition to it. Or yet again, if something has always existed, nothing must always have served as its substratum or receptacle, and is therefore eternally prior." This is our illusion. "A glass," we say, "may have always been full, but the liquid it contains nevertheless fills a void," though if it does it is merely a void in our consciousness, interest, or an unfulfilled desire. "In the same way" we falsely reason that "being may have always been there, but the nought which is filled, and, as it were, stopped up by it, pre-exists none the less, if not in fact at least in right." In my folly, "I cannot get rid of the idea that the full is an embroidery on the canvas of the void, that being is superimposed on nothing, and that in the idea of 'nothing' there is *less* than in that of 'something.'" Hence all the mystery."

Is not this in the main fine? For it means nothing less than that the supposed Nothing upon which all somethings are extended as on a carpet, is—far from being nothing— greater instead of less than they. And how beautifully it is implied that our finding of truth is wrapped up with our desire to find it and with our interest in it when found. But Bergson, like James, often sows wheat among his tares to give the crop a reputation. And what does he mean by suggesting that he has led us out of "mystery"?

He might have meant artificial mystery, man-contrived mystery, the mean mystery which comes of supposing that the greater comes out of the less. But the whole trend of his philosophy shows that he did not mean this, or not for long. He lapses into truth only when he forgets his principles. He is, in fact, a very ringleader among those who say that the less creates the greater. He differs from most of the others only in that he studied what is known as "living" matter rather than "dead"; was a naturalist, a biologist, rather than a physicist.

You may think that this is an advantage, but it so happens that the physicist is more apt to perceive the limitations of his science than is the biologist. He burrows into the roots of things, and finds the ground giving way beneath his feet. But the biologist, working in material that is midway, is the last man to see light from either above or below.

Added to this, Bergson was something of a mathematician, thus acquiring a tendency to be subjective. His protoplasm, his "flux," his God, his *"élan vital,"* his sole reality, maintains no permanent office in the Out There, still less in the Up There. Its address is 32 Rue Vital, Paris, and that is the only vital thing about it.

"For a mind which should follow purely and simply the thread of experience," he tells us—to go on with the chapter, "there would be no void, no nought, even the relative or partial, no possible negation. Such a mind would see facts succeed facts, states succeed states, things succeed things. It would live in the actual, and, if it were capable of judging, it would never affirm except the existence of the present."

Thus the cloven foot appears. He advises us here to give up thinking, and thus avoid the void. Were we to take the advice whole-heartedly, we certainly should avoid the question. We should cease to be aware of anything. To follow purely and simply the thread of experience would be to abandon all individual existence. We should lose ourselves in the flux. We shouldn't have any experience.

Practically, he may merely mean that it is best to have a good time, and not bother. But, as usual, his way of escaping from thought is to indulge in an orgy of false thinking. We are supposed to have rendered ourselves incapable of judging, and in this sad state we nevertheless pronounce judgment—that there is nothing but the present—precisely the sort of judgment one might expect. For obviously one couldn't be conscious of the present if incapable of comparing it with the past. And Bergson has so far lost his own judgment that he now denies that very Nothing which he has been at such pains to reveal as the great Everything-But-Ourselves. We are alone, and we cease to be. We become blind by dint of being all eye.

Of course it is not quite possible to make such a steep down-grade as this, but we can manage to go quite a distance in this direction. We cannot become completely absorbed in the passing moment, but we may become absorbed to a very great extent. Sometimes, when we give ourselves up to action, allowing no time for reflection, we lose almost all of our human attributes. We become raving animals. It cannot truly be said that "we" act at all. It is the body which is acting and reacting. This is what he calls "experience." But whose experience? Obviously, the body's.

Oh, never fear! It shall become ours later on. The body is quite capable of storing up experience for us. When the rut is over we awake, perhaps, to the symptoms of some terrible disease. In battle we are wounded, but do not feel it. But is it we who are wounded? Not yet.

It is merely the flesh. There are consequences in store, but we do not pay for them—until we pay for them. For a time we achieve a very fair degree of nothingness. We emulate, not altogether without hope, the thoughtlessness of the animal and the plant.

Nor is it necessary to go to all this trouble. If action abolishes thought for a time, so does a dream, a stupor. We can become just as blissfully stupid by mere dozing, with or without the aid of chemical soporifics. Action so absorbing as to obscure its purposes, may be good. Coma may be better. What we want to get rid of, it seems, is struggle. And one may get rid of struggle sometimes by plunging into a fight, sometimes by going to bed. Resistance gives the fight a moral quality by maintaining a view of an end to be achieved. Resistance keeps us awake. But surrender to anything brings peace—not necessarily the peace which passeth understanding. There is a peace which consists of failure to reach as far as the understanding.

Shall we then surrender to the unattainable, and be satisfied only with progress? Bergson recommends surrender to something nearer home. Let the body have its way with us. And then, if peace prove evanescent, we can at least maintain that it was not the dream that hurt, but the waking; not the fall, but the stopping so suddenly upon reaching the ground.

The practical flaw in his system is the fact that we always do wake up—though frequently only when it is too late to repair the damage. To lose ourselves in action is to lose the sense of distinction by forgetting the In Here. To lose ourselves in dreams is to lose the sense of distinction by forgetting the Out There. One always retains a vague shadow of the lost sense, but it may become very vague. All we can achieve is a more or less stunning blow upon the head—that troublesome head which harbors thought and prods us with awareness.

For thought reveals a moral situation, a self and others with which it is bound to deal. It even reveals Another, more formidable than our mere fellows. Bergson is a bludgeon, the great enemy of contemplation, of a brooding upon the Beyond. He has furnished the philosophical justification of the Chicago gunman on the one hand, and of the hop-head on the other. He and his like have made all sorts of irrationality respectable.

2. A CONTINUITY OF SHOOTING OUT

It must be said in Bergson's behalf that there were extenuating circumstances. Rationality itself had long since ceased to be altogether respectworthy—a loss of caste due to the antics of a younger and illegitimate brother, Rationalism, whose creed was and is "The mind is everything."

Now when a man says that the mind is everything, he of course means that *his* mind is everything. Or as Santayana puts it in *Winds of Doctrine*,[1] "Solipsism [the doctrine that nothing exists but the doctrinaire] has always been the evident implication of idealism; but the idealists, when confronted with this consequence [of their theories] . . . have never been troubled at heart by it, for at heart they accept it. . . . Idealists are wedded to solipsism irrevocably; and it is a happy marriage—only the name of the lady has to be changed."

The name of the lady has to be changed in most marriages, and as this particular lady is highly bigamous, if not ubiquitous, she has almost as many names as there

[1] New York: Charles Scribner's Sons, 1913, Ch. I, pp. 14-15.

are modern philosophers. It is heardly fair to charge her up-keep to the Idealists alone. Some of them, in fact, have quite other ideals. But the moment you attempt to construct all things out of one, you wed her irrevocably, whether you call yourself an Idealist or not; for there is *one* one which it is impossible to leave out, and that is yourself. The movement known as Modernism can be reduced to one principle, Self—and that is why it is a movement in unreality.

Solipsism's marriages all are barren. For far from being modern, the lady is probably as old as history, and—not being young with truth—is certainly beyond her bearing years. Nor are her marriages properly to be called marriages, since she and her putative husband are one—not in any softly impeaching metaphorical sense, but by hypothesis. Nuptials of this sort were formerly indicated by the term Absolute Immanence, from the Latin *in manere,* to remain in. And to remain in means not to go out. Anybody who tries to invent a philosophical system all his own is pretty certain to found it upon some personal fad, out of which all things are supposed to spring. But they have not the strength to spring out. If his name is Schopenhauer, he will call the big idea Will. If he is Nietzsche, he will call it Superman. If his name is Whitman, he will call it Walt. Bergson calls it Time, "real" Time. But of course in French—*la durée réelle.*

Here is a desperate attempt to get away from staying in without going out, for staying in is the negation of creativeness, to create being to send out. Bergson's Time, though *all* go-out, takes the stay-in with it, so to speak—which naturally prevents it from really going out. Time, like anything else, requires a not-self, a not-time in this case, to go out from.

There is a way of almost hiding the fact that an absolutely immanent single principle of any sort as humanly conceived is absolute nonsense. Take yourself, for example, and hold forth as follow:

"I myself posit the not-self as an idea." That sounds final, and as wise as an owl looks. It means—come on, let's pretend that it means something! It means: "I have more than one idea. There is my idea of myself, to begin with. If that were the only idea I had, I admit that all the unkind things you say about me would be true. But I also have lots of other ideas—my idea of you, to name only one. So we all exist, *I* as my idea and *you* as my idea. What do you mean, solipsism?"

I mean that, no matter how many things there are in the world, if they are all one and you are something, you must be It. If you, without any help, can send out a Not-Yourself which has an existence of its own, then you are God. But you cease to be absolutely immanent.

Life is not content to present us with one mystery, even though it be that transcendent mystery which is called deity. It presents us with another, whose proper name is always one which *means* soul. That God and soul are different is evident from the fact that we are living, sentient creatures. That they have something in common is equally evident for the same reason. The effect of our being either totally united with or totally separated from deity would be a total loss of everything we now call knowing or feeling, because what we call knowing or feeling is a mere comparison which we make between more or less of similarity, or (which is the same thing) more or less of dissimilarity.

Many are deceived into thinking they are Monists because of the love which they feel for this similarity, and do not know how to put into words. But philosophers ought to know how to put it into words. Instead, they try to put into words the unwordable. They either try to deny the Transcendent by bringing it within the reach of mind, or they try to force the mind to exceed its functions. They have what old Sam Johnson used to call, "a rage for saying something when there's nothing to be said." And so, many of us are inclined to leave philosophy to Scotchmen.

But why this mad desire to describe everything as a universal sameness, as unity in the form of an unfruitful eunuch? This passion for a ridiculous, false simplification? Is it a yearning for perfection? I doubt if that is always the case. It looks more like a very natural anxiety to escape from moral responsibility by assuming that perfection is already achieved. If we can only convince ourselves that all is essentially one, then it follows that good is essentially the same as evil, and we're all right whatever we do.

The impulse got a great boost at about the time of Henry VIII, when it was a comfort to many people to think that being outside the Church was the same as being inside, that Protestantism was essentially the same as Catholicism. They could then follow their king with a good conscience. Later, it was pleasant to be told that one remained essentially a Christian even when one ceased to be a Protestant save as a protester against Protestantism and Christianity and religion in general; that to disbelieve was the same as to believe; that the best way to observe morality was in the breach. This permitted one to preserve old emotional associations with the name of Christ, and to speak feelingly of the Carpenter while denouncing everything that the Carpenter stood for.

I could name a great many other reasons, as for instance:

(*a*) The prevalence of people who say it is a sin when we differ from *them,* who call no pleasures innocent except their own, who try to take the glorious dangers of freedom away from us.

(*b*) Our escape from a responsible authority only to fall into the jaws of an authority that owns no responsibility unless it be that of seeing to it that we are whatever the authorities would like us to be. The more dangerous among us are given the chance to be like them, the others to be like their servants. Really, these brother keepers of ours are a little trying.

(*c*) Our excessive democracy, which makes individuality perilous to all. In the United States, for example, a citizen finds himself in the minority of some hundred and ten million to one. If there is much diversity, the majority are likely to be united only by a desire to oppose the peculiarity of every single unit in the mass. It is thus that defiance of government in principle leads to the loss of individual freedom in practice. The surviving doctrine of the responsibility of the soul and its divine right of choice, has thus far saved us—to that degree in which we have been saved. If this should finally go, nothing will stand between us and the rule of demagogues, whose power lies in flattering the majority—which soon becomes an undistinguished and unworthy conglomeration, to which all but martyrs flock in self-defense.

But probably the chief cause of the growing unpopularity of the intellect is its own inordinate ambition in the recent past. Its failure to account for everything in terms of itself, becoming every day more notorious, is undermining its reputation. One votes it a dull dog. The fact

is, mind is of no use whatever except when supported by
moral considerations. It helps us to achieve ends, but
how can we achieve them or try to achieve them unless
we have chosen what ends to achieve? And ends bring
purpose into question; ends imply differences; ends show
that we love some things and hate others—and love and
hate are moral phenomena. Before we can think we must
have done something else much more fundamental. We
must have chosen. And the choice must be ours, not our
neighbor's.

Shall we never get over trying to use the mind to explain the essence of things? We can never hope to understand how reality came about, but perhaps we can eventually induce ourselves to believe that it did come about
—not half of it, but all of it—and that we are here, or
at any rate somewhere. Our brains will then assist us
mightily in carrying out what we have decided to do
about it.

Henri Louis Bergson, who is of Jewish extraction,
was born in Paris in 1859—just in time to arrive at
maturity when Modernism was in its dusty, and had not
yet entered upon its nasty, phase. It hardly called itself
Modernism then, but was divided into the opposing camps
of the Materialists and the Metaphysicians—Metaphysicians of a sort. These latter had reduced everything in
heaven and earth to concepts. The materialists (not for
an instant to be confused with the present and more
human if too sensual cult of the body), pent up in a
closed mechanical system by the genius of Newton, had
reduced to concepts everything in heaven and earth. There
was considerable hard feeling between Tweedledum and
Tweedledee. Science and philosophy still pretended to be
dealing with a living world, but it was beginning to be

felt that their subject-matters had died at the hands of analysis, and remained dead even when synthesis put a whole pile of the slain in a heap. Motion, save in an ellipse, as when a comet chases its tail, was scientifically taboo. Motion of any sort was philosophically taboo. The static was Mumbo Jumbo.

"Everything moves forward!" cried Bergson, in effect.

It was almost as if a fresh sea-breeze had suddenly blown across a desert.

Almost, but not quite. For if everything moves, then Motion becomes Absolutely Immanent, and—as is the habit of such Absolutes—ceases to move. A thing cannot move except in relation to something else which is relatively at rest. A breeze can't blow across a desert if it takes the desert with it. Not even a sand-storm would think of being quite so drastic.

But Bergson, being a literary genius of the first chop, was not troubled by the absurdity so often displayed in bald statements. He knew how to give bald statements a head of verbiage. It is dangerous to talk too recklessly about motion that can be seen. So let us turn to motion that can be felt. Duration! There is a subtile mover for you. Everything endures. And as he defines duration as change, everything endures by dint of not enduring. It is therefore duration which endures, change which changes not—because it keeps on changing! We are asked to think of time as something in itself, and without a background of timelessness. We are back to that function of a pump without any pump, whole cloth made out of a hole in the fabric. And with what gusto he spreads it before us!

"Duration is the continuous progress of the past which

gnaws into the future and which swells as it advances. And as the past grows without ceasing, so also there is no limit to its preservation. Memory is not a faculty of putting away recollections in a drawer or of enscribing them in a register. There is no register, no drawer; there is not even, properly speaking, a faculty. The piling up of the past upon the past goes on without relaxation. The past is preserved by itself, automatically [*sic!*]. All that we have felt, thought, and willed from our earliest infancy is there, leaning over the present which is about to join it, pressing against the portals of consciousness that would fain leave it outside.

"Memories, messengers from the unconscious, remind us of what we are dragging behind us unawares. We feel vaguely that our past remains present to us. What are we, in fact what is our character, if not the condensation of the history that we have lived from our birth—nay, even before our birth? It is with our entire past, including the original bent of our soul, that we desire, will and act. Our past, then, as a whole, is made manifest to us in its impulse; it is felt in the form of tendency, although a small part of it only is known in the form of idea.

"Thus our personality shoots, grows and ripens without ceasing. Each of its moments is something new added to what was before. We may go further: it is not only something new, but something unforeseeable. That which has never been perceived, and which is at the same time simple, is necessarily unforeseeable. To predict it would have been to produce it before it was produced. It is then right to say that what we do depends on what we are; but it is necessary to add also that we are, to a certain extent, what we do, and that we are creating ourselves continually." [2]

Thus, at the bidding of the wand of eloquence, Time becomes all but visible, a cloudy monster, swelling as it advances, gnawing into a future which has not yet been produced, perceived or foreseen—a thin diet, one would think. Yet the monster fattens upon it—"automatically," no doubt—presses against the portals of consciousness, and leans over the present with an air so positively threatening that it is no wonder the present scurries on, and (thank heaven!) manages to keep out of its jaws.

But this is not philosophy, this is a swollen metaphor. Memory ceases to be a *faculty,* and becomes a *messenger.* We are rid of a drawer, and acquire an errand boy. He comes from the unconscious, from the past. One would expect him to come laden with unconsciousness from such a source. But no, he comes laden with memories. He has changed the denomination of the securities with which he set out—a bold thing for a messenger who is not a faculty to do. Time is nothing but the As-Was. We feel its "drag" even before the boy arrives—and on what legs he moves Bergson's personal shooting, growing and ripening leave him no duration to say. I gather, anyway, that it is the drag that pushes us forwards into the future.

We are our past, without being anything to acquire any past, and this past includes the original bent of our souls, which evidently were bent by our past before we were past enough to have any. We must have created the bent continually while we were creating the past so as to

[2] *Creative Evolution,* by Henri Bergson, translated by Arthur Mitchell (New York: Henry Holt and Co., 1911), Ch. I, pp. 1–7. I have somewhat condensed the text.

give it a chance to create us. Could anything be more delightful than this construction of our present out of a past which was itself constructed out of future which was not?

Bergson is hard put to it, because he is seeking to have creation without any creative force. He has to have it, of course, but he hides it—now in a messenger boy; now in a past which has a drag which pushes; even now and then in us—or rather especially in us—who manage to pull the drag, although we are nothing but our own histories, our own past pulls, which now do pull the past. If it had not been for the inopportune arrival of an unconscious messenger boy disturbing the vagueness of my consciousness with an unconscious message from the unconscious, I might have believed it—for surely the less conscious one is the more reasonable this philosophy appears. But what is reason?

"The history of the evolution of life," Bergson tells us,[3] ". . . reveals . . . how the intellect has been formed by an uninterrupted progress along a line which ascends through the vertebrate series up to man. It shows us in the faculty of understanding an appendage of the faculty of acting."

So, though there is no faculty that remembers or helps us to remember, there is a faculty of acting and another of understanding. And the intellect was formed by a progress—i. e., it was formed by its own forming. And this took place along an ascending line, which was neither an ascent nor a line until after it had been ascended. Acting, not an actor, was the cause of mind. Acting made the actor; effect made the cause. And all this was revealed by "the history of the evolution of life," and re-

[3] *Ibid.*, Introduction, first paragraph.

vealed to "us," though we are ourselves, he tells us, a "condensation" of this very history. History must have been talking to itself in its sleep. And so understanding was shaken out of matter like dust out of a rag—a peculiar rag which had not dust in it until it was shaken out, which was not in fact a rag until it was shaken up. What makes this ascending matter? Creative evolution. In other words, shake shakes it, and the shake so shaken makes it. I hope no purist will be offended if I say that Bergson writes like an old hand at the bellows.

Now it might be thought that Shake, or action, is but a crude name for the acts of God—but Bergson shakes his head. God is the very last thing that is shaken into existence. He is the sum of all shakes, of all acting, the final appendage.

"The line of evolution that ends in man is not the only one. On other paths . . . other forms of consciousness have been developed, which have not been able to free themselves from external constraints or to regain control over themselves, as the human intellect has done, but which, none the less, also express something that is immanent and essential in the evolutionary movement." No, these are not God. But—

"Suppose these [were] . . . brought together and amalgamated with intellect," that is, with us? "Would not the result be consciousness as wide as life? And such a consciousness, turning around suddenly against the push of life which it feels behind," for the drag has now got to pushing, "would have a vision of life complete—would it not?—even though the vision were fleeting?" [4]

It would—not.

[4] *Ibid.,* p. xii.

"It will be said that, even so,"—even though things are as they are not—"we do not transcend our intellect, for it is still with our intellect, and through our intellect, that we see the other forms of consciousness. And this would be right if we were pure intellects, if there did not remain, around our conceptual and logical thought, a vague nebulosity, made of the very substance out of which has been formed the luminous nucleus that we call the intellect. Therein reside powers that are complementary to the understanding, powers of which we have only an indistinct feeling when we remain shut up in ourselves, but which will become clear and distinct when they perceive themselves at work, so to speak, in the evolution of nature. They will thus learn what sort of effort they must make to be intensified and expanded in the very direction of life." [5]

This sounds like a bid for the church vote. Bergson here distinctly gives up the notion that the mind is all, and at the same time seems to confess that it is something, a luminous nucleus, which has but to be extended in the direction of life in order to absorb more and more of the vague nebulosity which surrounds it. Yes, but in which direction does he look for life? Strangely enough, it is towards those "other forms of consciousness which have not been able to free themselves from external constraints or to regain control over themselves,"—a control which they appear to have lost without ever having had. We amalgamate their vague nebulosity with our own bright nucleus, and then turn around suddenly and look behind us, down the slope. Down there reside certain powers which become distinct the moment we plunge into the fog. So we perceive the whole of life, because

[5] *Ibid.,* pp. xii–xiii.

we are the apex, all is behind us and nothing before us. We make an effort to degenerate, lest we forget. Thus the vague nebulosity of the burdock and the snipe adds to our human luminosity by diminishing it.

I know exactly what he means, for I have often done it—and so have you. For have you never extended yourself in an easy chair, and "just relaxed"? An excellent habit, sometimes described as forty winks. And, just as you dozed off, have you not sometimes thought of those great things that you once meant to accomplish? And didn't everything become so nebulous that these things were magnified like street-lamps in a mist, until they seemed bigger than any works ever accomplished by man, and as good as done into the bargain? For the action that he talks so much about is not to be confounded with work, or with the putting of anything difficult into practice. He doesn't want us to think, but to feel—a feeling of large self-satisfaction. Not a bad thing, either. Nobody wants to be discouraged all the time, or worried, or on the rack. The mind certainly needs an occasional rest.

But it is not the Beatific Vision, or any vision. It is not a glad willingness to be small in the presence of greatness. It is more like an opium dream, or that which comes to those who are slightly intoxicated. And these bland emotions which drape the entrance to the Land of Nod, are, Bergson assures us, the stuff that Time is made of—real Time. Not the sort of time measured by "yonder clock," which you share with other people and use to help you keep your engagements. No. But "pure duration," out of which not only Time but everything else is constructed. He describes it somewhere as "nothing but a succession of qualitative changes, which meet into and permeate one another, without precise outlines, without any tendency to externalize themselves in relation to one another." Nor with relation to anything else, for that matter.

For this is the famous "flux." When we dream, "we no longer measure duration, but we feel it." Why not, then, abandon the luminous nucleus, and rejoice in the feeling of vagueness? Ah, but it so often feels bad. And its nebulosity, though it makes it feel bigger, makes it feel worse. For none of these drugs work permanently. They let us awake, or turn us over to nightmares, if too much indulged in. It might have been pleasanter could we have remained tulips and primroses, if such we ever were. But we can't get back. The "direction of life" is closed in that direction. We do not descend the ladder, even though we fall off. A degraded man is by no means as inspiring a sight as a pert and lively monkey, nor does he so much enjoy himself.[6]

But the stupor was evidently still working when Bergson went on to boast: [7]

"My mental state as it advances on the road of time is continually swelling with the duration which it accumulates: it goes on increasing—rolling upon itself, as a snowball on the snow."

Well, all I can say is that out our way snowballs do indeed increase when they roll on the snow, but we do not call this rolling upon *themselves*. Nor do we advance much on roads which lie behind us. It seems, though, that with a "mental state" it is different. This rolls forward, and finds itself all laid out ahead of itself—which troubles my luminous nucleus with a vague nebulosity. I confess ignorance as to the proper *modus operandi* of a thing

[6] See Appendix C.
[7] *Creative Evolution,* p. 2.

engaged in rolling upon itself. I have gone so far as to step upon my own foot, but that is all. However—

"Let us seek, in the depths of our experience, the point where we feel ourselves most intimately within our own life. It is into pure duration that we then plunge back, a duration in which the past, always moving on, is swelling unceasingly with a present which is absolutely new. But at the same time we feel the spring of our will strained to its utmost limit. We must, by a strong recoil of our personality on itself, gather up our past, which is slipping away, in order to thrust it, compact and undivided, into a present which it will create by entering. Rare indeed are the moments when we are self-possessed to this extent." [8]

I'll say they are! And as in the Bergsonian world we are this very past which we gather up, I am much afraid that we haven't gathered it, and therefore have not thrust it into the present that it will create; that we have in fact strained the will *beyond* its limit, and that consequently there isn't any present. Bergson seems strangely plausible in spite of it all, but this is because he had made us see a picture—something like a man throwing his cloak in the path ahead of him—a man, mind you, not the history of a man, however deeply that history may have marked him.

But "now let us relax the strain. Let us interrupt the effort to crowd as much as possible of the past into the present. If the relaxation were complete, there would no longer be either memory or will—which amounts to saying that, in fact, we never do fall into this absolute passivity, any more than we can make ourselves absolutely free. But, in the limit, we get a glimpse of an existence

[8] *Ibid.*, p. 200.

made of a present which recommences unceasingly—devoid of real duration, nothing but the instantaneous which dies and is born again endlessly. Is the existence of matter of this nature? Not altogether, for analysis resolves it into elementary vibrations, the shortest of which are of very slight duration, almost vanishing, but not nothing."

Thus does he destroy in a single sentence the objectivity of that flux upon which his reputation as a philosopher mainly rests. But evidently it is the flux of feeling which counts—though psychology tells us that there is no such thing, the smallest feeling being also "not nothing." So let matter be discontinuous if it likes. It seems to vibrate even when we relax, which relieves us of considerable responsibility.

"It may be presumed . . . that physical existence inclines in this second direction [towards relaxation], as psychical existence in the first,"—towards strain.

I seem to sense the ghost of a meaning. Bergson is haunted here by a philosophy much wiser than his own, the conviction that if we had no will we would have no life. This which he calls a strain, and foolishly describes as a plunge "back," as if we could escape from the present and find ourselves in the past, is really a recognition of that inner timelessness against which time is measured. The strain comes from a vain attempt to utter the unutterable. The will itself is more superficial than this inmost spring of the motionless from which it emerges creatively as assertive action, as a tone might emerge from silence. But we must not say that the Inmost is motionless, or silent. For it is imponderable, truly a Nothing in the sense of a Something-Other—other than all that may be said or thought. We only know it by learning that we do not know it, and we speak of it only to confess that nothing which we can know or comprehend in the least accounts for our being.

But Bergson uncomfortably suggests that it is our wills, our selves in action, which create this source, which he identifies with a past that is "moving on." Also it seems to be the present that is moving on—when we relax the strain. From this he passes to the consideration of "nebulæ in course of concentration," and finally to these amazing words:

"Now, if the same kind of action is going on everywhere, whether it is that which is unmaking itself or whether it is that which is striving to remake itself, I simply express this probable similitude when I speak of a center from which worlds shoot out like rockets in a fireworks display—provided, however, that I do not present this center as a thing, but as a continuity of shooting out. God, thus defined, has nothing of the already made; He is unceasing life, action, freedom. Creation so conceived, is not a mystery; we experience it in ourselves when we act freely." [9]

Unceasing life, action, freedom, then, are not mysteries! The making of the not-made does not puzzle Bergson. Why? Because, from what he here says, as well as from what has gone before, it is evident that he identifies himself with this "God." It is Bergson who experiences being this God, though he politely includes *us* by the purely grammatical gesture of a *"we."* What power he thus acquires of not considering himself a mystery even so, I am of course in no position to say. What *can* I say of Henri the All, who is a shooting out of continuity constantly constituted of his own continuosity?

Upon this high note he may be left, while we go on to consider the strange interlude which is Santayana, interpolated by capricious history in the swelling symphony of Relativity.

CHAPTER IV

A TILT WITH WINDMILLS

I. SANTAYANA

TO William James, this philosophy of Bergson's was like "the opening of new horizons." To George Santayana it was "a vegetative stupor." [1] As a critic Santayana goes at once to the head of the class. But is his own philosophy any more substantial than that of the Frenchman, whose descent into the "blooming buzzing confusion" of the flux of sensation he likened to an attempt to "share for the moment the siestas of plants"? Any more real than that of this American, whom he described and damned as "credulous"? I am afraid not. It is much more subtle, much more beautifully expressed, the fruit of a wider culture, that is all.

And yet, dreading the *descensus Averno* which this book I am writing imposed upon me, I was at first tempted to take George Santayana as my guide and counselor to play the Virgin to my faltering Dante. He has, for one thing, some experience of the route, since his most alluring volume, *Dialogues in Limbo*, lays its action in that dim but not too uncomfortable corner of the Old Pagan Quarter indicated by its title. For another, his words have that musical persuasiveness which makes nonsense seem plausible and self-contradiction a

[1] *Winds of Doctrine*, Ch. III, p. 80.

— GEORGE SANTAYANA —
He built air-castles and then pumped out the air

mere modulation, as from a major to a minor key. Surely no more delightful companion could be found with whom to saunter through the less torturing portions of modern hell. That he could get me in, I felt certain, notwithstanding his predilection for antique dress. But could he get me out?

I was not so sure. Or rather, I was sure he couldn't. He has never been able to get himself out, in spite of his sly mocking of all others who are in. How he hates the new-fangled Bottomless Pit! And yet it isn't so different from the ancient Hades in which he proudly strives for citizenship, though the early decorations are better.

Santayana was born in Madrid in 1863, and has a Catholic mind—I mean a Roman Catholic mind. His interest is in wholes, not in parts. And though he passed through Harvard and then went to England to pursue post-graduate studies at King's College, Cambridge, he has never succeeded in adopting, or even in understanding, the Protestant point of view. His experiences at Harvard —he returned there in 1889 as Professor of Philosophy, and continued his connection for twenty-two years, with an interlude in 1905–6 as lecturer at the Paris Sorbonne, —seem not to have impressed him favorably. Afterwards, he refused a professorship at Oxford, shook the last of Reformation dust from his feet, and retired to what may be described as a little monastery of his own at Rome —where he lives to-day, like a hermit.

Or perhaps I should say, like a poet—for intellectually he owns to no religion. Poet he always has been, and his very first book, published in 1896, was entitled *The Sense of Beauty*. Since then he has published several volumes of verse, including *The Hermit of Carmel* (1901). Here speaks an exiled Spaniard—of the Renaissance. But the exiled *thinker* goes—or tries to go—back to the days before Christ. Self-deprived of the Church, not liking her enemies, he seeks a philosophy that includes at least all the wit and wisdom of its day. Hence his passion for the classics. But can a man live as if two thousand years had not been? He cannot.

And so, in Limbo, he describes himself as The Stranger, and tells the shades who cannot be his companions:

"I have been a stranger in all my dwelling-places." [2]

"Had there been," he goes on,[3] "some wise prophet in my day, summoning mankind to an ordered and noble life, I should gladly have followed him, not having myself the gift of leadership. I should not have asked him for the absolute truth nor for an earthly paradise; I should have been content with a placid monastery dedicated to study, or with a camp of comrades in the desert. But I found no master. Those who beat the drum or rang the church-bell in my time were unhappy creatures, trying to deceive themselves."

"I confess that sometimes the fair vision [of appearances] intercepts my reason, and the passers-by point at me in derision for standing amazed at nothing in particular and seeing gods in the commonest creatures." [4] "But there is a lusty core in the human animal that survives all revolutions; and when the conflagration is past, I seem to see the young hunters with their dogs, camping among the ruins." [5] For "after all, I am a child of my time." [6]

[2] *Dialogues in Limbo* (London and New York, 1925), p. 24.
[3] *Ibid.*, p. 33.
[4] *Ibid.*, p. 28.
[5] *Ibid.*, p. 85.
[6] *Ibid.*, p. 25.

Not to be at home anywhere; to stare amazed at nothing in particular; to see God nowhere but in the lusty core of the human animal is to be a child of the times indeed. Santayana, the classicist, is but the neo-neo-pagan, the ultra-modernist turning back upon himself. These lusty young hunters with their dogs are but a part of the blooming buzzing confusion of Bergson, the "thick" of the later James. And, beautiful as they are, they are very thin—a romantic picture of action drawn by one who seldom stirs from his garden. Real hunters with real hunting-dogs, their muzzles stained with the blood of little birds, would fill Santayana with horror.

And not even these lusty sportsmen are really at home in the present. Nobody is, for the present is strange and new. The very dogs grow wistful from their association with men. To-day is always a ruin—the dear remains of that far, unfallen yesterday of which we still continue to dream. No man builds a home for himself, but for his children. His own home was his father's house. If that falls, life is never the same again—as anyone who has lost so much as an earthly parent, can tell you. But there are such things as wilful orphans, spiritual waifs by election, who try to live lustily among the dogs, pawing among the rubbish of alien civilizations or even the alien elements of their own—stones which will not be made bread—while behind their backs the ancestral hall and all its store still rises sweetly in the sun. Have we really no prophets save those engaged in the chase?

No man was ever less fitted than Santayana to be happy in a camp. All his loves twine about the supernatural, the eternal. But the modern philosopher is forbidden to have faith. His very faith in sense begins to fail—hence the wild scramble to catch at feeling ere it be gone. He is a slave to a presupposed cipher, to an iconoclasm that breaks the idols of dogma—as if dogmas were anything but data, things "given," without which nobody can or does either think or live. He must keep on terms with dead intellectualist humbugs without a future, or with the jejune imaginings of young hunters without a past, and all the latest fads of anything that calls itself science. Religion is barred upon pain of loss of academic standing, and with it that meekness, or willingness to learn everything, which alone can inherit even the earth.

I do not for a moment suggest that the academician is any more a hypocrite than are the rest of us, but merely that he has his loyalties. Santayana has more of these than he thinks. He began, like his confrères, by giving the mind a trial. But his earliest important work, though called *The Life of Reason,* confesses itself in the preface to its second edition to have been but a history of "the murmur of nature, wayward and narcotic." This is precisely the direction in which the rest of his world was turning at about that time. It dates.

In this same Preface, written in 1922, twenty years after the first appearance of the book, he declares: "There has been no change in my deliberate doctrine; only some changes of mental habit. I now dwell by preference on other perspectives. . . . What lay before in the background—nature—has come forward, and the life of reason, which then held the center of the stage, has receded." Not much of a change, even of mental habit, for a historian of a reason that is a murmur of nature, to let nature take the center of the stage—though it must be admitted that nature murmurs more in the later comment than in the book itself.

"The vicissitudes of human belief [now] absorb me

less," the Preface continues. "The life of reason has become in my eyes a decidedly episodical thing, polyglot, interrupted, insecure." What, then, is secure? Not anything, unless it be the wayward narcotic.

"When our architecture is too pretentious, before we have set the cross of the spire the foundations are apt to give way." [7] Exactly—if the cross is to be on the spire only, with no bit of lignum crucis—that *"più lieve legno"* of which Dante sings [8]—slipped beneath the corner stone. Santayana, like all other Unrealists, hesitates to slip anything beneath anything. "Hypostasizing," he calls it.

In *Platonism and the Spiritual Life,* he criticizes Plato for hypostasizing morals, that is, for making the knowledge of good and evil the basis of philosophy. He complained of the early Bertrand Russell (the later Russell has got bravely over any such fault) for a similar breach of faithlessness. And yet Santayana himself was finally compelled to do a little hypostasizing on his own account, and he chose—that same old "murmur of nature, wayward and narcotic," which he discovered to be among the things hypostasized for him in the first place, though eventually he was to name it Animal Faith, and write a book about it.

Animals, it seems, have faith. They believe what they see, hear, touch, taste and smell. Man might well do as much, at least until something which is neither smell, taste, touch, hearing nor bodily vision teaches him not to rely too implicity upon his five senses. I doubt if even dogs are utterly lost in the flux or lack altogether the

[7] *The Life of Reason,* (New York: Charles Scribner's Sons, 1922). Preface to this, the second edition, pp. v–vi.
[8] *Inferno,* III, 93.

wits that go with young hunters on two legs. But before we go into this, let us note what Santayana has to say about faith of a purely human sort—for unto Santayana it was given to define Modernism as "an ambiguous and unstable thing . . . the love of all Christianity in those who perceive that it is a fable. . . . The heroic attachment to his church of a Catholic who has discovered that he is a pagan." [9]

He is speaking more particularly of modernists whose bringing up has been literally Catholic, who have remained Catholics in sentiment but not in belief. It is they who are said to assume that Christianity is a fable—though how it could be more of a fable than the fables that have striven to take its place, I cannot imagine. The definition seems perfectly to describe its author. Also it is capable of a much wider application than the one he has chosen to give it. For though the brains of Modernism bear the marks of the circumcision rather than of the Cross, the modernist outside of Israel, even when Protestant in immediate background, may well be described as a Catholic who has lost his faith. We may call ourselves pagan, but as the Reformation was confessedly an attempt to solve a problem in subtraction rather than in addition, we are fated to love the old Roman remnants, or nothing. If we have lost any faith, it was a faith in something that came to us through Rome, no matter what we called it.

But I fail to see anything particularly heroic even in the conduct of those unbelieving Catholics; for either their attachment must have come unattached, or else has proved too strong for such courage as they have in their new convictions. But their nostalgia can be understood.

[9] *The Winds of Doctrine*, Ch. II, p. 49.

It is the same homesickness that weighs down, in a lesser degree, perhaps, the unbelieving Protestant. That culture which surrounded one at childhood, presenting itself like the mother's breast and receiving the heart's first and inalienable affections, is not to be supplanted by another quite outside of it.

Our so-called knowledge of the Greeks, in so far as it was not acquired coldly through printed matter, was told us in play as so much mythology. But do you think real pagans were taught by their mothers that Pallas and Zeus were fictions, symbolizing it may be some facts in the scenery of Hellas but otherwise meaning nothing? No; to be pagans we would have to worship the gods. And we don't. As Ludwig Lewisohn makes Reb Moshe say in *The Island Within*, "The Way is not communicated by any report of any book, but from soul to soul."

True, in the main at least. And so, though Santayana has of late listened chiefly to the murmur of that nature which speaks Greek, he is less a pagan than he is an unbelieving medievalist, peculiar chiefly in that he knows what words were used during the Middle Ages—knows them as Avicenna [10] knew the fourteen books of the *Metaphysics* of Aristotle, "perfectly, by heart, both forwards and backwards," all that escaped him being "how the doctrine could be true."

But even this peculiarity—some slight awareness that there *was* a period following Plato and Aristotle and preceding Bacon and Descartes—gives Santayana the air of a quaint, almost legendary figure in this our Modernist world. How the light from this horizon tinges even what he writes of to-day! Listen to what he says of the Vatican:

[10] *Dialogues in Limbo,* The Secret of Aristotle, p. 175.

"Many an isolated fanatic or evangelical missionary in the slums shows a greater resemblance to the apostles in his outer situation than the Pope does. But what mind-healer or revivalist nowadays preaches the doom of the natural world and its vanity, or the reversal of animal values, or the blessedness of poverty and chastity, or the inferiority of natural human bonds, or a contempt for lay philosophy? Yet in his palace full of pagan marbles the Pope actually preaches all this. It is here, and certainly not among the modernists, that the gospel is still believed." [11]

"The modernists talk a great deal of development, and they do not see that what they detest in the church is a perfect development of its original essence; that monachism, scholasticism, Jesuitism, ultra-montanism, and Vaticanism are all thoroughly apostolic; beneath the overtones imposed by a series of ages they give out the full and exact note of the New Testament." [12]

So the Modernist, whose modernism consisted, a few pages back, in his "love of all Christianity," now finds something to detest in the Catholic Church—its Christianity, to wit. Paradoxical, but true, as a man may hate that chastity in an adored woman which keeps her from being his. Nor can it be denied that the Modernist feels a genuine love for the new-found freedom—i. e., bonds—of his separation. He does not lay all of the blame upon himself. Probably none of it. He argues that if the spouse had been perfect, it need never have happened. She might have been more understanding. Perhaps she might. She had and has her human and errant side. Mingled with her virtue there may have been mere pruderies, as a little

[11] *Winds of Doctrine,* p. 53.
[12] *Ibid.,* p. 52.

virtue mingles with the vice of her harlot successor. He
would like to return home if he could take his Mimi with
him, and live in broad-minded, polygamous comfort. But
even Mimi refuses. She also is exacting. As for revivalists,
I doubt if Santayana ever met one in the flesh, or he would
know that our Billy Sundays are quite apt to be Funda-
mentalists, by no means adverse to preaching the doom
of the natural world or the reversal of animal values.
In being just to the Vatican one need not slander the
soap-box.

Apart from sentiment, Santayana will have traffic with
neither. And though he does not hurl at the Church the
usual accusation that its bells are cracked with age, his
objection goes deeper—that they hypostasize something!
Thus, although at times he seems inclined to take a hand
at the bell-rope himself, it is never to call to worship any
more than it is to sound an alarm. He merely loves the
music, while refusing to hear the unplayed tune that it
strives to *mean*. The summoning of mankind "to an ordered and noble life" goes steadily on, but he—having
hypostasized self-deceiving summoners—has decided to
conduct his own self-deception without assistance.

"Seldom does any soul live through a single and lovely
summer in its native garden, suffered and content to
bloom," is his lament.[13] But he intends to be the exception.
"My own philosophy," he continues, "I venture to think
is well-knit in the . . . sense . . . [of] being conceived in
some moment of wonderful unanimity or of fortunate
isolation . . . My eclecticism is not helpless before sundry
influences; it is detachment and firmness in taking each
thing simply for what it is." And so he retires to his ivory

[13] *The Realm of Essence* (London and New York, 1928). Preface,
p. xviii.

tower, *"un nobile castello, sette volte cerchiato d'alte mura,"* literally a hotel suite, which ironic chance has located almost within sound of the loudly hypostasizing bell of St. Peter's.

But, as he permits Democritus to remark in the elder Limbo, "blooming is not knowing, and roses and cabbages should not be founders of sects." To take each thing "simply for what it is," may mean to take it simply for what it isn't. Does he not himself declare, through the mouth of his Alcibiades, that "illusion may be truly pleasing while we think it true: but to cling to it knowing it to be illusion is ignominious and well nigh impossible"?

For says this Democritus: "The dreamer can know no truth, not even about his dream, except by waking out of it. . . . Beauty is a fleeting appearance . . . the Stranger . . . knows that it is an appearance. . . . He nevertheless cherishes it more than reality." And "what is your present plight? Dispersion and impotence of soul." [14]

This, I take it, is the dreamer's own verdict passed upon himself. I can't see anything else it means. His animal faith seems not to have saved him. But what is animal faith?

2. ROSINANTE

"Animal Faith," as described in the book called *Scepticism and Animal Faith,* represents Sanatayana's revolt from "the verbiage of metaphysics," from words that "pass as decent drapery for our ignorance," even though they may be but an "eloquent expression of it." [1] Here the "murmur of nature, wayward and narcotic," already heard in *The Life of Reason,* swells to an uproar, a nar-

[14] *Dialogues in Limbo,* p. 25 *et seq.*
[1] *Realm of Essence,* p. 179.

cotic uproar be it understood. But lest I be quite deafened and put to sleep, I shall content myself with following its echoes in the author's other works. I wish to show that it is the obscure motif from which all his themes are developed.

For Santayana is a gentle sensualist of the imagination, forever writing one movement or another of a Pastoral Symphony, with its little pipings, its little storms; its little river—unlike Beethoven's—wandering through blooming fields of the poppy. "Blooming is not knowing." We can never know. Therefore let us bloom. This is the meaning of the whole refrain.

It is inevitable that a poet should seek for something real, something concrete, something which he can hold and love. And if he be the sort of a poet who can make his Democritus sigh in Limbo, "All human philosophy, except science reckoning without images, is but madness systematic, putting on a long face," what is he to do? Santayana knows little and cares less for science of any sort. He merely pays it lip service now and then. What he wants is a series of images about which he can weave the garlands of poetically sad emotion.

"The fruit of my experience is that I despise rhetoricians and demagogues and moralizers and comedians, and respect rather the rough arts and passions of mariners and soldiers, the patience of ploughmen, and the shrewdness of merchants or of the masters of any craft; all people acquainted with danger and hardship and knowing something well, though it be a small matter, and each striking out bravely, like an honest bird creature, to have his will in the world." [2] Thus might Don Quixote, weary of the attempt to prove that windmills

[2] *Dialogues in Limbo,* pp. 82–83.

are nothing but ogreish ideas, turn to admiring contemplation of Sancho Panza at his chores about the barn.

"The flight of eagles and the swimming of porpoises," says the Stranger,[3] "are admirable to me in the realm of truth: I rejoice that there are such things in the world." Who does not? "But I am not tempted to experiment in those directions." It would be too much like work. In imagination, however, it is rather pleasant to pretend. And then it becomes "so simple to exist, to be what one is [or is not] for no reason, to engulf all questions and answers in the rush of being that sustains them."[4] Just what he ridiculed James and Bergson for maintaining.

It isn't altogether a bad move. Better be an honest blind creature than nothing. But it seems to me that the Don has now passed Sancho Panza, and handed over the reins to his horse. Well, if all human philosophy is really but madness systematic pulling a long face, I for one am all for the stable. I shorten my long face with a horse laugh. Rosinante, hear my prayer! Thou believest that when thou perceivest something, there is something which makes thee perceive it. Teach me thy animal faith—for even a man may attain to it, I understand, if he will but stop stopping to think. Neigh, the world does move—from the manger at Bethlehem to the manger de la Mancha. We stretch ourselves out in the thick flux of the straw on the floor of the box-stall, murmuring narcotically, "Oh, what's the use!"

"The 'sane' response to nature is by action only."[5] This is the action—to lie down. And immediately it becomes

---

[3] *Dialogues in Limbo,* p. 173.
[4] *Realm of Essence,* p. xiv.
[5] *Ibid.,* p. ix.

exceedingly agreeable to contemplate exertion—in others.

It was perhaps a mistake to despise the comedians, for comedy comes from a sense of proportion. When the humorist closes one eye it is not because he cannot see out of it, but to indicate how comic it would be if he couldn't. Or wouldn't. Not the honest but the dishonest blind are his game. Why must we always look at this famous "action" with a squint which *really* conceals half of it from view? There are many ways of acting besides waving your arms and legs, or even wagging your tail. Turning our backs upon the horse and the ass, there is the cow, who at least chews the cud. One can ruminate without leaving the blessed barnyard. Why, action might even include the practice of religion—said to be an amazing help to the profession of it—and justify such old-fashioned and highly scientific texts as, "O, taste and see how gracious the Lord is!" Is pudding the only thing which we ought to put to "pragmatic" proof? Isn't it a bit droll always to apply the principles of Behaviorism to rats in mazes, and never to saints in glory, or on the way to it?

Anyway, how are we going to respond sanely to nature unless we interpose a little reflection between seeing and doing? Blind creatures may be honest, but they do not have their will in the world. They do not have any will. Blind reaction is not will, it is the absence of it. Nor is reaction faith, even of an animal sort. Faith implies belief in something, it therefore cannot be blind. When we call it blind we deny its very nature, which is to see. Where there is no perception there is no belief. It is impossible to accept anything upon no evidence whatever. We believe there is a far side to the moon, but wouldn't were it not for the testimony of the near side. Faith is the ability to see evidence when presented, and is thus the opposite of credulity, or the gentle art of ignoring evidence. It demands a complete and humble surrender of our natural unwillingness to learn, hence its exceeding unpopularity. It has the further peculiarity of never being deceived. Therefore perfect faith has been called the *scire recte,* true vision, or Right Reason.[6]

But for right reasoning to arrive at complete results, all the evidence must be in and available. It is not all in for any of us, and much of what is in is unavailable for most. Vision of any sort is a gift, like an ear for music. If you are tone-deaf, not all the pleadings of Palestrina can make you hear a tune. Do you therefore conclude that the motet, *"Exultate Deo adiutori nostro,"* is not music? That depends, of course, upon whether you are capable of accepting the testimony of others. Possibly you are ready to believe that when someone is moved to tears by listening to a jumble of sounds, he hears something that you do not. You accept his word, having found it dependable in the past in regard to things which you yourself are capable of putting to the test.

In this indirect fashion, the least faithful of us supplement our shortcomings a thousand times a day. I find Brazil in a cup of coffee, and India or China or Japan in a cup of tea, though as yet I have visited none of these countries. In this way I have come to believe in the existence even of saints, and to reject the theory that they are merely people who have fits. Their history seems to me different from that of epileptics.

When we are mistaken, it is because we have been blind to some of the evidence, or have wilfully shut our eyes to it. Mistakes come from ignorance, or from ill will. You can't always blame a man for making mistakes. You

[6] See Appendix D.

can't blame a toad for not reading music. Refusal to cultivate a musical faculty, is another matter. But to speak of deceived faith is to use terms that contradict each other, like wrong Right Reason, or saying that we do not have the experiences which we do have. For the basis of faith is experience, that is to say, dogma.

The modernist hates the word dogma. And yet, what is a dogma? A musician will tell you that there is melody in the Eroica. He will be quite dogmatic about it. Why? Because he hears it. There is no use in arguing with him. He is dealing with what he knows, and is positive. To him the melody is a datum, something given by experience. So, I repeat, all dogmas are data, and the very word dogma means something given. I don't see how we are going to get away from them, though it must be admitted that the word has a bad name, and "dogmatic" is supposed to apply to all positive assertions, especially those unsupported by anything but vocal energy.

The dogmas of religion seem to be of another sort from those of everyday secular life, but are they? Not according to the claims of religion itself. They may or may not be ours, but that is no reason why we should go on ignoring what they profess to be. It is ridiculous to oppose them because they are dogmatic. The belief of Miss Helen Wills that there is such a thing as a tennis ball, is also dogmatic. It is a faithful deduction that she has drawn from the evidence of her senses. Those to whom the doctrine of the Incarnation, for example, is a dogma, must also deduce it from the evidence of their senses—by rightly applying reason to data. Otherwise it is not a dogma, but an assertion only. It follows that if this doctrine is indeed a dogma there must be, in some men and women, a spiritual sense supplementing the familiar five

—a spiritual sense more developed or natively more keen than it is in unbelievers, or enjoying better opportunities. This is a question of fact upon which nobody in whom this sense is less developed or instructed is competent to pass judgment. His ignorance is, in the language of theology, invincible—that is, it is not to be overcome by any act of his will. He is, by hypothesis, in the predicament of an English sparrow confronted with Blackstone's *Commentaries*.

Christianity is often said to be illogical and unreasonable, though the slightest acquaintance with its higher forms shows it to be as logical and reasonable as astronomy, to say the least. But its premises are admittedly hidden from the naked eye. As to the existence of telescopes, there is considerable evidence accessible even to those who have never looked through one. It will hardly make them astronomers, but there it is.

We have wandered a long way from animal faith, however. Must we go back? It seems so, though I can't quite see why we should ignore the fact that we are human. Animals we are, no doubt, and we also have vegetative processes within us, and chemical processes within them—hardly a reason for adopting the point of view of sulphurated hydrogen. Nevertheless, an interview with the cow proves immediately interesting, for—

"Knowledge such as animal life requires," Santayana tells us,[7] "is something transitive,"—i. e., something having an object. It is "a form of belief in things absent or eventual or somehow more than the state of the animal knowing them." Good! She does better then than some who neither part the hoof nor chew the cud, but merely chew the rag. "It [knowledge such as animal life re-

[7] *The Realm of Essence*, p. 2.

quires] needs to be informative. Otherwise the animal would be the prisoner of its dreams, and no better informed than a stone about its environment, its past, or its destiny."

There is no gainsaying it. The stone is a prisoner of its dreams, or to its lack of dreams. The cow wakes up to the discovery of calves and fodder as things absent, or eventual, or somehow other than cow. For "the dreamer can know no truth, not even about his dream, except by awakening out of it," says Democritus to the Stranger in Limbo.[8] But [9] "the heart of nature is full of dreams; and I daresay there is a poet in every nut and in every berry. But the soul of animals must be watchful; they cannot live on mere hope, fortitude, and endurance."

Nuts and berries dream on. As Miltons they must be mute and inglorious. I think we flatter them when we say it is *their* song we hear, no matter how sweetly something may sing through them. Why praise the piano wires when it is DePachmann who plays? But, to leave these lowly, instrumental regions, "They [the animals] must hasten to meet perils and opportunities, and dreams are fatal to them. Action being necessary, a true perception is indispensable."

How far we have come from that contented dog whose master was "out"! It now appears that cows think. I never doubted it, myself, having been brought up on a farm. Their true perception amounts to apperception, mental synthesis, the Herbartian union of part of a new idea to one already in the mind. They make deductions and discover resemblances, and differences between this and that. They hasten to meet not only their perils but

[8] *Dialogues in Limbo*, p. 27.
[9] *Ibid.*, p. 77.

their opportunities. Blooded stock, evidently. Their faith, you see, is just like ours. They are not quite "engulfed in the rush of being." They know their beans.

Does that make them less poets than are the nuts? I should hate to think so. Less easily amenable as poetic subjects, no doubt, for these busy, practical, four-footed folk are more difficult to mould to the forms of our fancy. But I disparage this disparagement of hooves in favor of roots; this suggestion that the poet is a business-man on his way to become a pansy; this confusion of con-templation and creative imagination with the dullness of vegetables euphemistically described as "dreams." If we must emulate the quadruped, even, let us do it by know-ing our own beans.

"Your philosophy," my dear Santayana, if I may quote the rejoinder made by your Democritus to your Alcibi-ades,[10] "would be perfect if, instead of being a king, you had been a cabbage." But, as Democritus says again,[11] "roses and cabbages should not be founders of sects." Why, then, should quadrupeds? They are good as far as they go. I like your animals immensely. They show that the *narcotic* murmur of nature is not in their lowing, but in our own ears. But are there really no better prophets, even in your day, summoning mankind to an ordered and noble life?

"Without intending it," perhaps, "you confirm the doctrine of the divine Plato, that the liver is the seat of inspiration." [12]

As Democritus replies to Dionysius the Younger, once Tyrant of Syracuse, "I rejoice . . . all the more that, if that saying, as usual, was inspired, it was not the politic

[10] *Loc. cit.*
[11] *Op. cit.*, p. 9.
[12] *Ibid.*, p. 9.

Plato who uttered it in his waking mind, but his honestly dreaming liver that uttered it for him." From a liver, it was just the sort of an utterance that one might expect.

Now let it not be thought that I am insensible to the ungracious nature of my task, even of a sort of cruel absurdity in trying to catch all these wingèd words and hold them to strict account, and so robbing a beautiful night-moth of its congenial dusk merely to break it upon the wheel. But how else get at the thought which lurks beneath the soft Santayanan shadows? Nor is it a passing mood that is here being dissected. The exaltation of the animal, the vegetable and the mineral is constant, and he is but lightly disguised as "the Stranger" when he declares:

"Of all men I am the last to belittle the world of matter or to condemn it. I feel towards it the most unfeigned reverence and piety . . . For I know that matter, the oldest of beings, is the most fertile, the most profound, the most mysterious; for it begets everything, and cannot be begotten; but it is proper to spirit to be begotten of all other things by their harmonies, and to beget nothing in its turn." [13]

Not to "belittle" matter is a mild way of putting it when one is about to hail it as the begetter of spirit, even this barren spirit—which is really not barren even though it means here only emotion. And now the renowned liver begets thought.

For "as the privilege of matter is to beget life, so the joy of life is to beget intellect; if it fails in that, it fails in being anything but a vain torment." [14] Thought, then, is matter's grandchild only, born of the hepatic "Life"; and Life's sister, "Spirit," remains childless, thus re-

[13] *Ibid.*, p. 174.
[14] *Ibid.*, p. 170.

sembling its intellectual niece, since "this intellect ought to be sterile," and probably is, "because it is an end and not a means." So he compares it to the lyre, which "has performed its task when it has given forth the harmony, and the harmony, being divine, has no task to perform. In sounding and in floating into eternal silence, it has lent life and beauty to its parent world."

How did the "divine" get into this family? Or was it the divine matter of the lyre that touched the matter of the strings? But evidently Santayana is only playing with us—or to us—on *his* lyre.

"Be not deceived by the language which philosophers must needs borrow from poets, since the poets are the fathers of speech," and thus the fathers of liars.[15] Moreover, this *Limbo* from which I have been quoting is confessedly a work of fiction, contrived delightfully in the manner of the Socratic dialogues of Plato. It reveals its author's predilections none the less. He pretends that "nothing . . . has shaken in the least" his "old allegiance" to Democritus, i. e., to science.[16] But "if he salutes the atoms from a distance, it is only in condescension to the exigencies of art . . . He honors reality only for illusion's sake, and studies in nature only pageants and perspectives, and the frail enchantments that are the food of love." [17]

If music be the food of love, play on. But this is an Alfred de Musset rather than a Shakespeare among the pundits. His thought is not thought, only emotion. We feel the stir of the body—emotion in its more lowly stretches. These lovely flowers of his summer garden,

[15] *Ibid.*, p. 192.
[16] *Ibid.*, p. 85.
[17] *Ibid.*, p. 88.

pregnant with the spirit of *dolce far niente,* which he calls "action," are of a gorgeous color but with a breath of cloying sweetness, slightly poisonous. The weather is *scirocciale.* The *tramontana,* or cool north wind of cogitation, has ceased to blow. There is rather a *tramonto,* a setting, a going down, a decline, what Italians call a *tramortimento,* or swoon, the lazy motion of instincts faintly sexual but all too lethargic to be up and doing ere they yield to insensibility and slumber. This intellect, which is an end and not a means, is sterile in fact. This incense, which pretends to be offered up to nature, is mixed with ether. Nature snorts and runs away from it on four nimble legs. It is grateful only to the nostrils of sick Man, bent on getting enjoyment out of the dulled pains of his malady. Heaven has become a hospital of wilful incurables next door to the morgue. Here is spiritual defeatism.

We slander the animals if we think we lie down with *them* on such beds of asphodel. "The soul of animals must be watchful. . . . They must hasten to meet perils and opportunities, and dreams are fatal to them." So they are to us—such dreams as these.

3. DON QUIXOTE

But a highly cultured gentleman like Santayana could not remain satisfied with a faith enjoyed in common with animals, either natural or perverse. He would have been glad to keep on sharing their belief in corn and the windmills that grind it, but like his illustrious countryman, Don Quixote, he really wanted to get at whatever it may be that lurks behind the mills and makes them go. So his next book was called *The Realm of Essence.*

It is a promising title, especially as its author once

said that "the essence of anything is simply the whole of it," thus reminding us that whenever we break up anything into its parts we immediately cease to deal with that thing and begin to deal with other and lesser things—as when we analyze a tune into its individual notes, not one of which, we shall find, has, individually, any tune whatever. The essence of truth, then, is all truth. It looks as if the wind of the spirit were about to blow, and the mills of God to turn. We seem about to supplement the animal with the human, and then be introduced to Cause, to reality in the round.

But it was not to be. We are to continue the modernist game of leaping from one incompleteness to another. Our bread is still to be all top, all bottom, or all middle. For the essences with which we now have to deal must not be confused with things which are essential. They are essentially non-essential. We are therefore advised to begin by losing even our animal faith, and thus unburdened to set out along the way of doubt.

"The sceptic, once on this scent [of doubt] will soon trace essence to its lair. He will drop, as dubious and unwarranted, the belief in a past, an environment, or a destiny. He will dismiss all thought of any truth to be discovered or any mind engaged in that egregious chase. He will honestly confine himself to noting the features of the passing apparition. At first he may assume that he can survey the passage and transformation of his dreams," —that is, he may assume that at least he can see an appearance; "but soon, if he is truly sceptical and candid, he will confess that this alleged order of appearances and this extended experience are themselves only dreamt of, like the future or the remoter past or the material environment—those idols of his dogmatic days." [1]

After having been at some pains to discard the idols of *my* dogmatic days, I find that Santayana never wrote this book of Essence at all, for it is a part of my material environment. I dreamt it. A truly candid sceptic would go further, and confess that he cannot survey even his dreams. The survey also is a dream. So he dreams that he dreams that he dreams that he dreams. "Nothing will remain but some appearance now; and that which appears, when all gratuitous implications of a world beyond [in the sense of a world outside], or of a self here, are discarded, will be an essence."

Do you follow? We drop all gratuitous implications of a world. There appears to be an appearance left. So we drop the implications of a self, of that to which the appearance appeared. Everything which implies either a self here, or an environment there, goes simultaneously by the board. They are *out*. That which still appears, after the last remainder ceases to remain and the last appearance to appear, will be an essence.

"Nor will his own spirit, or spirit absolute [of the candid one] be anything but another name for the absolute phantom, the unmeaning presence, into which knowledge will have collapsed."

Collapsed appears to be a well-chosen word, though no word whatever is empty enough so much as vaguely to shadow forth this emptiness. An essence, then, is what is left of anything when it is entirely gone. It is the presence of absence. Santayana attempts to go Bergson one better and to encompass the Idea of Nothing—not as the pure other, now of this, now of that, but as the

[1] *The Realm of Essence*, p. 2.

All's Pure Other. He tries to think of everything as non-existing all at once—and of course the best way of doing this is not to think. He invokes the vague emotion of an idea.

"Essences," he tells us,[2] "do not exist." Why then write a book about them? Because "essence so understood more truly *is* than any substance or any experience or any event."[3] One sees now what he is driving at. Essence is neither ourselves nor matter. It is Being as distinguished from Existence. It is that which lies in the mind of God before creation, *ante rem* as the Scholastics express it. And sure enough we hear him saying:[4]

"My position is simply the orthodox Scholastic one in respect to pure logic, but freed from Platonic cosmology and from any tendency to psychologism." From any tendency, that is, to admit that souls and bodies were ever among the things created,—that individuals actually exist. This is almost equivalent to saying that nothing was ever created at all, that the universe remains *ante rem,* still undistinguished from its source. Is even the remainder "scholastic"? No, because of a perverse inversion, everything goes wrong.

"The multitude of essences is absolutely infinite."[5] Though undifferentiated, they seem to be numerous. "It is an accident to essence to be manifested; but not to be manifested is also an accident; it means simply that matter or intellect happen never to have traversed that form."[6] You are asked to think of an essence as an

[2] *Ibid.,* p. 21.
[3] *Ibid.,* p. 23.
[4] *Ibid.,* p. 93.
[5] *Ibid.,* p. 21.
[6] *Ibid.,* p. 32.

empty frame, which matter or intellect may happen to
make manifest by entering and filling up.

"The realm of essence . . . since it is immutable and
incapable of any local emphasis or arbitrary existence,
can have no influence on the production of anything." [7]
He has made the creature produce the Creator, which
would lead to the supposition that the *post rem* produced
the *in re* and the *in re* the *ante rem*—and he calls it
"orthodox"! He attempts to foist upon the Scholastics
the doctrine that the ineffable Principle back of the Creative Mystery has no influence upon anything! What, then,
has? Matter! That appearance which appears to a dreaming, non-existent self, which so lately disappeared! "As
the privilege of matter is to beget life, so the joy of life
is to beget intellect. . . . Matter, the oldest of beings,
is the most fertile, the most profound, the most mysterious; for it begets everything, and cannot be begotten." [8]

True, these latter remarks, here requoted, are dropped
by Santayana's personal shade in a fancied Limbo, and
it would be wildly unfair to hold him responsible for
every whim expressed in the ghostly badinage that he
makes the chief delight of Shadowland. But he reiterates
the idea time and again in his other works.

"The realm of essence . . . is an infinite field for
selection." It is a field of possibilities without power.
For these Santayanan essences do not give power. They
give to nothing its possibilities. Things seem to own their
own possibilities, to be their own real essences. What,
then, is this other essence? Does it mean anything to
you? It doesn't to me, though I can see the idea that is
at the back of Santayana's mind and that he denies by

[7] *Ibid.*, p. 179.
[8] *Dialogues in Limbo*, pp. 170–174.

refusing to grant that it stands for something with force, for creative originality, even that most mysterious of all which bequeaths the power of creative originality to another. But he would locate this power only in the recipient of this power, as if the legatee enriched the ancestor. And so he adds:

"To appeal to fact, to bump existence with empirical conviction, is . . . but to emphasize some essence, like a virtuous bridegroom renouncing all others. . . . But the bride after all is only one of a million, and the mind has simply wedded an essence." Then it is the mind, itself created by privileged matter, that is in turn privileged to create the fact to which it unites itself in marriage! But essence "cannot select or emphasize [or bump] any part of itself." [9] "An essence is an inert theme, something that cannot bring itself forward, but must be chosen, if chosen, by some external agent; the choice made by this agent [is] wholly arbitrary." [10] Conduct, I would say, wholly unbecoming an "agent."

Essence "is an invitation to the dance," but it does not dance unless invited by the band-master, and it leaves it to the band-master to call the tunes—which I hope are not too inert. "When the selection takes place, we accordingly refer it to a different principle, which we may call chance, or fact, or matter. . . . The selective principle," *le chef d'orchestre*, "is matter."

Matter, which we lost along with our animal faith, we—now that we have lost it—install as the thumping bridegroom in a happy harem where none can thump back. Scepticism, in silencing matter's clamor, has somehow made it the big noise. Matter alone selects the es-

[9] *The Realm of Essence*, p. 15.
[10] *Ibid.*, p. 20.

sences that are to be emphasized into existence by being slapped upon their wedded backs—and at that it would seem that they must be slapped before they can be there to be married. Moreover, matter is a *principle,* and it answers to the name of Chance. It begins to look to me as if chance were a hypostasized force whose hypocephalic hugger-mugger is responsible for all the hircine hurry-scurry of this our hebetude.

More marvelous still, "this world of free expression, this drift of sensations, passions, and ideas, perpetually kindled and fading in the light of consciousness," is "the Realm of Spirit." [11] Matter itself thus chances to de-materialize, to become the drift of sensations that chance to be its own effects. and these effects, so expressive of their freedom from their own causes, are spiritual! Essences are the golden drones of chaos. Nobody works in our house but my unessential old man!

And so, our bread which tried to be all top, flops over and becomes all bottom. Idealism, grasping at nothing, loses its hold and falls into gross materialism.

"The subjective attitude in philosophy," says Santayana, "is not only prevalent in these times, but always legitimate; because a mind capable of self-consciousness is always free to reduce all things to its own view of them. . . . A scrupulous scepticism, falling back on immediate appearance, is . . . a chief means of discovering the pervasive presence of essence." [12]

We have now, evidently, the essence of that once-despised Solipsism. The subjective attitude is of course legitimate, even inevitable, if by that is meant the habit of looking through one's own eyes. But to whom does

[11] *Ibid.,* pp. x–xi.
[12] *Ibid.,* p. I.·

it give the right to reduce all things to his own view of
them? Wouldn't it be better (after the viewing) through
animal faith to infer the landscape? I wonder how even a
scrupulous scepticism is going to fall back on appearence,
immediate or not, when the sceptical and candid sceptic
is himself only another name for an absolute phantom?
As a non-appearance falling back upon non-appearances,
I personally have some difficulty in discovering the essence of their pervasiveness.

But I do feel the thrill of an unacknowledged and mysterious Presence, which is possibly all that Santayana
intended. Not even in the midst of what purports to be
cold reasoning, can he forbear his rôle of poet. But, unlike music, poetry—especially philosophical poetry—has
a running bass of thoughts, and these should be harmonious and valid. Such poetry has been. Santayana, however,
seldom escapes from the hypnotism of his own phrases.
He keeps forgetting what he has previously said, as if
it were of no account. Do you remember our faithful and
believing cow? She, too, now grows doubtful.

"Intuition, or absolute apprehension without media
or doubt, is proper to spirit pursuing essences; it is impossible to animals confronting facts." [13] It is then,
animal faith that doubts; and spirit, which was but late
the spirit of doubt, pursues essences without doubt, and
apprehends them *without media*—a considerable task, I
should say, even for spirit, and especially for that sort of
spirit which it is the privilege of matter to produce.

And what is this "intuition" he speaks of? "Intuition
is thought." [14] And yet "the organ of intuition is an

[13] *Ibid.*, p. xi.
[14] *Ibid.*, p. 67.

animal psyche governed by the laws of material life, in other words, by habit." [15]

Spirit, then, pursues essences by virtue of an animal psyche governed by the laws that govern matter; and matter is governed by habit. And as habit is something that arises only after certain things have been done in a certain way for a considerable time, the conduct of matter is governed by something coming later than the conduct. I do wonder what governs habit, especially as it seems to be habit that governs chance. Moreover, this precious "realm of matter" must be "begged separately," —apart, that is, from the realm of essence. So both are begged, and neither is given. We do not begin by finding ourselves anywhere. We are governed by the "habit" of begging. And in a pinch we will even beg the question.[16]

As if to add to the confusion, Santayana speaks of "the imaginary as a whole, conceived as the realm of essence";[17] of "the most interesting essences, like the thoughts of ancient philosophers"; of "complex essences, like Euclidian space"; of "a simple essence, like Pure Being."[18] This is not adhering even to his own definition of essence, for he cannot mean that the imaginary is the cause of nothing, when even an imaginary bear will cause us to run. Nor can he wish to infer that the thoughts of the ancient philosophers have had no influence. He must have intended to say that the unimagined-imaginary, and the thoughts of the philosophers before they were thought, were essences. Power came to their empty glory

[15] *Ibid.,* p. 37.
[16] See Appendix D.
[17] *The Realm of Essence,* p. 30.
[18] *Ibid.,* pp. 68–69.

when they were traversed by matter, or by matter's off-spring, mind! However—

"Of all essences, the most lauded and the most despised, the most intently studied in some quarters and the most misunderstood in others, is Pure Being. It has been identified with nothing, with matter, and with God." [19] I feel no inclination to identify it with God, myself—not the sort of Pure Being that we have here. Santayana would make the "essence" of God a sort of sheer blankness which somehow, though powerless in itself, gives to matter the power of emphasizing the infinitely numerous divisions of its undividedness. We are now asked to think not only of our unthought thoughts, but of the forceless possibility of thoughts and things unthought and unknown to God himself.

"Even if some philosopher or some god thought himself omniscient," Santayana assures us,[20] "surprises might be in store for him, and thoughts new to his thought; nay, even supposing that he was not a victim of sheer egotism in asserting that nothing more could ever exist." That is, a god thinking himself omniscient and justified in thinking so, might assert that nothing more could ever exist, and be surprised to find new things coming into existence, nevertheless and notwithstanding. And as matter has all power and might and initiative, it would of course be matter which brought these surprises about.

This "Pure Being" is Pure Bergson—squeezing a god out of the flux. And as the flux has already been reduced to appearance, God is created by the individual to whom they appear. And as this individual has also been eliminated a dozen times, we arrive at pure Unrealism, the

19 *Ibid.*, p. 45.
20 *Ibid.*, p. 21.

Void—though how we arrive at it, since we are *non est,*
I leave for you to explain. I myself give up even trying
to accept an Omniscient God, who is not omniscient and
exists by virtue of a perpetual rain of surprises coming
out of something considerably less than my own surprise
at the suggestion. We are in bocardo.

At the same time, all this is neither here nor there. It
was Santayana himself who taught me not to depend
entirely upon my reason in judging an argument, but also
to use my nose. As the shade of Democritus, in the first
Dialogue in Limbo, so well remarks:

When . . . the soul issues from the eyes or lips tur-
bid and clotted by virtue of the distorted imprints which
she bears of all surrounding things, she also stinks; and
she stinks diversely according to the various errors which
her rotten constitution has imposed upon her. Hence,
though it be a delicate matter and not accomplished without training, it is possible for a practiced nose to distinguish the precise quality of a philosopher [I presume
he means philosophy] by his [its?] peculiar odor, just
as a hound by the mere scent can tell a fox from a
bear." [21]

Well, what I have been smelling for several pages of
quotation is the reek of midnight oil offered as incense
to one musty human concept after another set up each
in its niche and pretending to be the God that it seeks to
hide. The musk of individuality is all very well, and
philosophers should doubtless be poets, men who have
recreated their learning so that it raises no dust from
dead mummies of truths that have given up the ghost
in being acquired. But they need a clearing-house of
some sort when they turn into philosophers. For how

[21] *Dialogues in Limbo,* p. 4.

shall the individually warped and distorted mind be straightened out and made just without some correcting contacts? I love, I must admit, the sweet incense that rises from the totality of human achievement.

Time was when each philosopher, though working in his own little cell, belonged also to a brotherhood. He served an apprenticeship and profited by the long results of many devoted to a common aim, a common faith, an unselfish love of truth. The world's masterpieces, even of literature, have come not infrequently out of such circumstances as these.

For in one sense, all good things are orthodox, since true orthodoxy is nothing but that just proportion which comes from knowing all that can be known about a subject. If you cut yourself off from any portion of relevant knowledge, you become a heretic, and your individuality consists largely in the mistakes you make. You may achieve fame from the same source, but not even your mistakes will be original. The world has wagged so long now that I fear all heretical "essences" have, ages ago, been traversed by thought. The heretic is usually right in spots, and a useful irritant. But if he be really heretical, and not merely a new hand holding the old torch so high that the timid fail to recognize it, he is but a half man, astigmatic with vertigo. Orthodoxy alone is truly inventive, for it alone has continually to rediscover and point out the unaltering worth in the midst of ever changing currents of appearance. It requires more talent and ingenuity to stay at home and keep one's house in order than it does to elope with every fancy that would like a nest built for it in the wilderness.

Santayana built air castles, and then pumped out the air. He tried to imagine an oracular tripod with less than three legs, or with no legs at all; a vertebrate without backbone. He sought to doubt that broad foundation of the inexplicable but undeniable upon which all logic must rest, and ended by thinking that he had made bricks without either straw or clay. I doubt if even bell-ringers could deceive themselves as gorgeously as he has done. But let us be grateful for the sad splendor of his failure. Besides, he is no more than half deceived. He does not take himself seriously. "I have ever loved philosophers," he says in Limbo, "overlooking and pardoning the foolish doctrines which they chose to profess, since necessity and custom compelled them to profess something." [22] Such candor is disarming.

I doubt that Aristotle really thought that "the soul of the world" is a "habit of matter" which leads it "now towards one form and now towards another," and that this habit of change is the "one true cause" of change, "the only principle of genesis anywhere." And I regret that it should be rumored, even in Limbo, that " 'Tis love that makes the world go round, and not, as idolatrous people imagine, the object of love"; that "the object of love is passive and perhaps imaginary." [23] But I agree that "there is no more bewitching moment in childhood than when the boy to whom someone is slyly propounding some absurdity, suddenly looks up and smiles. The brat," says Santayana,[24] "has understood. A thin deception was being practiced on him."

Just so. And this bewitching moment comes to me now. Someone has been slyly propounding not one absurdity but a multitude of absurdities. This brat has understood. Yet he is tempted to yield for a moment to the

[22] *Ibid.*, p. 75.
[23] *Ibid.*, p. 186.
[24] *The Realm of Essence*, p. xix.

gust of feeling that blows across these pages like a gentle, enervating breeze across the Roman campagna from the south. For feeling's sake our very thoughts are hushed and dreamy. For feeling's sake we even try to banish them altogether. I should have liked the walk better, myself, were the guide less given to going through the motions of climbing a pretended stair while stubbornly refusing to set foot upon a real one. But at least he has escaped that dreadful classroom manner which afflicts so many of his contemporaries. He is a "venerable and courtly man," whose *Limbo,* if no other book, is like the "crimson damask" it should be wrapped in, and worthy of "the silver clasp with its black opal" which should complete its binding.

"Let those excel who can in their rare vocations [of saint and philosopher]," he begs us, "and leave me in peace to cultivate my own garden. . . . I frankly cleave to the Greeks . . . and I aspire to be a rational animal rather than a pure spirit." [25]

Very well. Let us leave him. We shall go further and fare worse.

[22] *Ibid.,* p. 75.

CHAPTER V

RELATIVELY SPEAKING

I. EINSTEIN

I HAVE often thought that most of us would be happier and better off if we had a little more wholesome and total ignorance, and not so much false knowledge clogging what we are pleased to call our minds. But nowadays one would need to be deaf, dumb, blind, and live in a cellar, in order to retain any firm hold upon an innocent lack of misinformation. The "dark ages," when it was generally admitted that a mystery was something which could not be explained, are over and gone. For mystery has been substituted mystification, and the gentle art of fooling yourself. To be great in this art, so that you can fool not only yourself but others, you can do no better than to learn the language of the Higher Mathematics.

Now I have nothing to say against the last of the three R's. Long live mathematics, high, middle and low! At the same time, there is no denying that a mathematician is an Unrealist—though sometimes of an innocuous and non-poisonous sort. He deals with abstract conceptions, notions, whose abiding place is within himself. Whether these are based upon anything which may be perceived or even imagined, is far from his immediate concern.

If you ask him what price brass tacks when they come

fifty in a box and cost twice as much as unhatched chickens, with eggs at thirty cents a dozen, he will tell you that, apart from questions of freight, breakage, backache and interest on capital, brass tacks in your frame of space-time are a bare $2.50 a box. But he cannot, as a mathematician and upon the data here given, guarantee that the eggs are fresh, that the tacks are sharp-pointed, that the dollar is at par, or even that there are any tacks or eggs in the market—still less dollars in your pocket. He takes his data from the physicist, the poultryman, the hardware dealer, from Tom, Dick and Harry, and he says to them:

"If what you say is true, my answer to the problem is true also. If the answer is wrong, there must have been some mistake in your observations. Myself, I don't make mistakes." Hence the atrocious proverb, "Figures don't lie."

Figures always lie. They don't invent falsehoods, but they pass them on, magnified enormously by the process of logical deduction from false premises. They are the seven-league boots that enable one who begins a journey in a slightly wrong direction to beat all pedestrian records of going astray. It isn't true, for instance, that two plus two equals four—not in the world that we perceive. It isn't true of eggs, for example. Nor is it true, in the world of omelets, that one-half plus one-half equals one whole. You can't split an egg exactly in halves, still less can you take the half of one egg, add it to the half of another, and have an egg necessarily equal to any particular egg. There are eggs and eggs, and one egg differs from another in glory. But sometimes it is convenient to make believe that all eggs are alike. The mathematician takes that make-believe and juggles with it, per-

~ ALBERT EINSTEIN ~
With a stroke of the pen he brought the earth to a halt

forming a number of operations none of which are possible—with eggs. But he isn't dealing with eggs. He is dealing with make-believe, with concepts, with abstractions. The world of mathematics is a world simplified for human convenience. It is purely subjective.

Those simple lines without breadth, surfaces without thickness, points without body, by means of which the undergraduate tries to cross the Bridge of Asses—they do not exist in nature, and any conclusions based upon the supposition that they do are necessarily misleading. When we try to think of a surface without thickness, what we we really do is merely to leave thickness out of consideration, dropping from the calculation any results that might happen because of thickness. We can't imagine a straight line, but we can imagine a straight stick—though never shall we meet with one. It is easy to deal with these ideal sticks, always cutting out any reference to their breadth.

In lower mathematics, the kind we do our shopping with—and heaven knows they are high enough—we always start with things that are imaginable. We don't try to picture what they would be like if the features we ignore were actually non-existent. We simply ignore them as of no immediate consequence—as a judge who knows something when off the Bench may behave as if he didn't when he is on. In lower mathematics when the operation is over, we can check the result by experiment, and frequently do. Little harm is done, few bones are broken, and we generally remain safe to be left at large.

Higher Mathematics is a more dangerous game. We begin with abstractions derived from experiences of an unusual sort, usually in the way of physics or astronomy —for it will never do to forget that before a mathematician sets to work, somebody else (or perhaps the same man in a different capacity) has been busy noting what is going on and noting it slightly amiss, and at the same time failing to note a great deal that is going on beyond his ken. His errors are of course carried into the problem, disguised as facts. They do not influence the logic, but they have a mighty influence upon the conclusion.

So long as this can be tested by further observations and experiments, everything is lovely and the goose hangs high. Sometimes, however, the goose hangs too high. It may happen that the "facts" were obtained from a philosopher, a poet, or a madman. They may be profoundly true, and then again they may not. And experiment may be out of the question. A mathematician is not compelled to deal with abstractions derived from correct data, or even imaginable data. The unimaginable, the inconceivable, the self-contradictory, will do just as well. The Higher Mathematics therefore always tends to become the Higher Humbug.

I do not mean that the differential calculus, the integral calculus, or even the tensor calculus of Relativity, are humbugs. On the contrary they are, regarded as processes, among the highest achievements of human ingenuity. But they give no guarantee of the correctness of fundamental assumptions. Mathematicians have been clever enough to invent a language that outsiders do not understand, but learning a language does not confer the gift of wisdom and of good judgment. The language of mathematics rather tempts one to cut loose. It is easy to invent a sign to stand for anything, even the ineffable, and then it becomes easy to forget that, once introduced into an equation, it cannot be canceled save by another ineffable. A figure that looks much like an 8 lying on its side is, for example, the mathematical sign for the Eternal. With that in an equation, beware of attempting to translate the solution in comprehensible terms!

We have for many years been living under a tyranny. Newton proved with figures that the universe is a self-winding clock, composed of an unchanging amount of matter and an unchanging amount of force. Though personally a very pious man, he made it seem irrational to speak either of God or of free-will. It is now generally admitted that his system was founded upon slightly erroneous observations and several untenable premises. But we escaped from the frying-pan only to get into the fire. Came—not the dawn but Einstein and Relativity. Not to swallow him whole is now to incur the charge of weak-mindedness, though he restores our will only to threaten our wits. Who is he, then? And what is Relativity?

The "who" question is easy to answer. Albert Einstein is a German Jew—born at Ulm, Württemberg, in 1879—with an intelligent twinkle in his eye. He now wears a short mustache over a pleasant, dreamy smile, and possesses a remarkable musical temperament. He plays the violin, like certain other well-known Dictators, making up the music as he goes along.

As a schoolboy, he lived in Munich, attending the Gymnasium there until he was sixteen, when his father and mother moved to Milan. But Albert went to Switzerland. The Technical High School of Zurich busied itself from 1899 to 1900 in teaching him all those physical and mathematical truths that it was *then* idiotic not to accept. Between the years 1902 and 1909, however, after he had become a naturalized Swiss citizen and while he was holding a sinecure as engineer in the Swiss Patent Office, he elaborated some highly original truths of his own. For it was in this Patent Office that he wrote a little paper for the German *Annals of Physics*—it dealt with the effects of motion on measuring-rods, and appeared in 1905—which was not only a paper but a bomb.

It exploded, and the theoretical edifice of Nineteenth Century Science, founded upon the physical stability of rods, came tumbling down. Samson hurled aside but the pillars of the Philistine temple roof—the modern thought of his day. Einstein removes even the cellar. Modern thought collapses. If ancient thought remains, it is only because it rests upon quite a different basis. It isn't Einstein's fault if thought of any kind has a single leg left to stand on.

The *Annals of Physics* paper was followed by a number of yet more explosive books, concluding with the recent magnum opus, *On a Unitary Field Theory*. But the gist of his doctrine has been neatly embodied in a work that in English bears the title, *Relativity, the Special and General Theory*.[1]

Zurich, at first stunned, as was the rest of the world, to hear that time was the fourth dimension of space (it had been so described in whispers before); and that space, though having no real size, was crooked, and shaped like a battered tomato-can, woke up in 1909 and made Einstein Professor Extraordinarius at its University. Two years later, Bohemia secured him for the University of Prague. Another short interval, and Zurich won again—this time the Polytechnicum being the specific

[1] Translated by Robert W. Lawson and published by Henry Holt & Co., New York. It is the only volume of Einstein's from which I shall quote, and all references are to the edition of 1920.

beneficiary. But in 1914, when Germany was supposed
to be thinking only of war, the Prussian Academy of
Sciences in Berlin appointed him to succeed Van't Hoff.
Berlin's darling he has remained ever since. It does not
sound much like a tale of hardship.

It is, however, the tale of a very simple life, and it
has its amusing side. For instance, only a few months
ago the City of Berlin decided that an adopted son who
had won so much honor must be relieved of the last
vestiges of wordly care, and presented him with a hand-
some villa near Neu-Cladow, on the Havel River. But
alas! The City Fathers had never studied Relativity, nor
realized the far-reaching consequences of the precept set
forth in *Die Grundlagen der allgemeinen Relativitäts-
theorie* of 1916: [2]

"Every reference-body (coördinate system) has its
own particular time."

So when Frau Einstein (she is her husband's second
wife and his cousin), using her own reference-body and
coördinate system, thought it was time to take possession,
she was met at the villa's door by a representative of
Frau von Brandis, the property's previous owner.

"According to von Brandis time," said this wretched,
four-dimensional creature, "the time for you to take
possession will not arrive in this co-ordinate system for
five years yet."

What is worse, there were legal, three-dimensional
documents to prove it. Therefore, relatively speaking, the
Einsteins will have to sleep in the open until about the
year 1935.

Meanwhile I like to think of that young Einstein, who
was not only a Samson but a new sort of Elia, busy

[2] *Vide, Relativity,* p. 32.

with his own visions and his own twinkle in the unsuspecting Patent Office, as another Elia was busy in an equally unsuspecting East Indian House gaily burning down imaginary buildings for the sake of roast pig. It makes one wonder. What, do you suppose, is the clerk who sells you stamps at the local Post Office really thinking about?

2. THE MOLLUSK

Einstein was thinking about light, and the curious fact that it always seems to travel at the rate of 299,776 kilometers a second. If you are in an express train, and a crow is flying along the right-of-way, you will pass the bird less rapidly if it is going in your direction than you will if it is going in the opposite direction. But a light-ray will make the same speed in relation to the moving train, regardless of direction. Whether the ray and train move in the same or opposite directions, makes no difference. And this speed is the same that the ray will make in relation to the motionless railway embankment. It seems unreasonable. So for a long time it was supposed that there was a difference, though too small to be detected experimentally. Few express trains go even the hundredth part of a kilometer in a second—not fast enough either to add or subtract from the apparent velocity of light in any observable measure.

It so happens, however, that, according to current astronomy, the earth itself is virtually an express train rushing along its orbit at the rate of 30 kilometers a second. Light coming towards us from a star lying ahead of us (curvature of the earth's orbit is too trifling to matter) ought to pass at the rate of 299,796 plus 30, or 299,826 kilometers a second; while the light coming from another star and overtaking us from the rear, so to speak, should pass at 299,796 minus 30, or 299,766 kilometers a second. We have instruments capable of detecting a difference a tenth as great as this. Yet they stubbornly refuse to detect any difference whatever. Evidently something is wrong.

Nobody cared to lay the blame to light; so a mathematician by the name of FitzGerald filed a complaint against our measuring-rods. All bodies that are in motion, he said, are shortened in the direction of their motion. Their atoms crowd together. All earthly measuring-rods are in motion. Therefore, when we lay them along the line of that motion they are shorter than they ought to be, or than they would be if laid at right angles to it. A light-ray overtakes the earth from behind at the rate of 299,796 kilometers, it is true, but they are [1] abbreviated kilometers, shrunken with their own motion. There is no way of actually catching them at it, because whatever we try to measure them with proceeds to shrink likewise. And this universal shrinking of matter in motion came to be called "the FitzGerald contraction." It was held to be proven by the impossibility of its being discovered.

There was, however, a way of discovering it—in the opinion of H. A. Lorentz. If rods in their atoms are in truth packed together by motion, there would be certain electrical phenomena; and these, too, remained conspicuous by their absence. So he improved FitzGerald's theory by supposing that the atoms themselves, nay, the very electrons and protons of which they are composed, suffer

[1] It would make no difference in the case of the bird were the train at rest in reference to the air—as when a forward wind is blowing.

contraction. This explained how rods could shorten without any inner crowding, and explained the absence of any electrical demonstrations of protest. The FitzGerald contraction was renamed "the Lorentz transformation." It accounted beautifully for the apparent speed of light moving in our direction, when the problem is why it seems to go so fast; but it failed to account at all for the speed of light moving in a contrary direction, when the problem is why it goes no faster. Few people seem to have noted this objection, but there it was.

And now Einstein took the field. "Gentlemen, you don't know what you are talking about," he said—or polite words to the same effect. "You tell me that the earth is moving some thirty kilometers along its orbit every second. Applesauce! You mean, I suppose, that it is moving at that rate in relation to some imaginary fixed point in its orbit. I am not from Missouri, but I am—which is considerably more to the purpose—from Ulm. I should like to be shown that point. Kindly mark it with a piece of chalk. Drive a nail into it, or send a boy out there to hold one end of the tape-measure. You can't? Then to Hades with your point! What has an imaginary point got to do with the speed of light? Look here! I'll measure the velocity of light in relation to the earth—that's the real point. It goes 299,796 kilometers a second whether it comes from the east or from the west, from the north or the south, whether up hill or down. Why shouldn't it? It moves neither *contrary* to the direction of the earth's motion, nor *in* the direction of the earth's motion, for the *ganz und gar gut* reason that the earth hasn't any motion."

With a word to the Lord, Joshua made the sun stand still upon Gibeon and the moon in the valley of Ajalon.

With the stroke of a pen, Einstein brought the earth to a halt.

A stationary earth is not a new idea, nor one altogether without merit. The story of Joshua has been ridiculed for ages on account of its "impossibility." The Einstein miracle has met with almost universal acceptance and applause—such is the enlightenment of the age in which we live. Ptolemy would be at home again could he return now to an earth once more established where he left it. Come to think, was there a not a bit of trouble when Copernicus, and more particularly Galileo, upset all literature by the mention of some nonsense about a central sun? Did not certain ecclesiastical gentlemen suggest caution, and even invite Galileo to remain a guest at the Palazzo di Medici, in Rome, until he was willing to admit that a theory was only a theory? I believe so. And how they were laughed at for their pains! For years their conduct has been cited as a smoking example of bigotry. But it begins to look as if he laughs best who laughs last.

After all, the earth *is* the center, for all practical purposes, and if the famous "proofs" that she goes a-gadding are found to prove nothing of the sort, I for one shall offer no objection. But it will be said that I have put my own silly words into Einstein's mouth, and that —either through ignorance or because of malice—I have grossly misrepresented the master's point of view. I have, in fact, done something worse. I have translated the theory of Relativity into language that at last a human being may understand.

Einstein does not, of course, announce flatly that the earth stands still. Theoretically, he is very liberal, and quite willing to assume that the central honor belongs to the sun, the moon, some asteroid or comet, or what you will. He goes further, and declares that in any case it is only an assumption, that nothing can be said really to stand still or really to move. This is too much. Clearly it cannot be true of any material universe that actually exists outside of the mind of the beholder. Einstein is the very prince of unrealists.

It has been said that he does not pretend to be a philosopher. And yet his theories claim to embrace the very ends of heaven. They involve all of consciousness. If Philosophy finds no room within them, where shall she set up her house?

To assume that nothing is either at rest or in motion is to assume that nothing is, save pure and unimaginable Ens, or Being. But of course when we try to think of such a thing we merely play a trick upon ourselves. If we can find no fixed point without, we look for it within. So, practically, Relativity offers us the choice of two alternatives. Either the earth, and more particularly that point upon which the observer stands, is really and truly at rest —in which case the conclusions formed upon the experiences of everyday life are valid, or else there is no discoverable point of rest, and all points are equally good as locations for observatories for the reason that all are equally bad. From which state of affairs Relativity deduces its one general law of nature. viz., *that everything which happens, happens to happen in the way it does happen, and not otherwise.*

After due consideration, I am not inclined to dispute it. This is Einstein's famous Law of Gravitation, which Professor Eddington expresses thus: "Every body continues in its state of rest or uniform motion in a straight line, except in so far as it doesn't." [2]

[2] *The Nature of the Physical World:* New York: The Macmillan Co., 1929, p. 124.

Now there is a difference between saying that no point of rest is discoverable and saying that no point of rest exists. In the first case we assume that we are somewhere, even if we don't exactly know where. But in the second case we assume that there *is* nowhere. All distances, speeds, shapes, states of rest or of motion become theoretically impossible. Therefore nothing either moves or stands still. Nothing has any shape, and there is no near or far. This is pure non-existence, and reduces us to the images of our own imagination, which certainly do have shapes of some sort. The theory of universal relativity, if regarded as a fact, effectively abolishes the physical world and abolishes all real communication between man and man. All arguments against Solipsism apply here, and are sufficient to refute the idea that this sort of Relativity is true.

It will be objected that Einstein has nevertheless made several definite discoveries. Exactly. *But he has made them from the earth.* That's how he was fixed. One of these was an actual discovery—that light-rays are bent toward any material body which they pass, or as Einstein himself phrases it, "In general, rays of light are propagated curvilinearly in gravitational fields," [3] the word "general" meaning that this is a law of general or universal application.

"The curvature . . . required by the general theory of relativity is only exceedingly small for the gravitational fields at our disposal in practice. Its estimated magnitude for light rays passing the sun at grazing incidence is nevertheless 1.7 of an arc. This ought to manifest itself in the following way. As seen from the earth, certain fixed stars appear to be in the neighbourhood

[3] *Relativity,* p. 88.

of the sun, and are thus capable of observation during a total eclipse of the sun. At such times, these stars ought to appear to be displaced outwards from the sun by an amount indicated above, as compared with their apparent position in the sky when the sun is situated at another part of the heavens." [4]

This, when first written, was prophecy. But its truth was triumphantly demonstrated by star photographs taken during the solar eclipse of May 29, 1919.

The other discoveries were in the nature of new explanations. It has long been known that the lines in the spectrum are shifted when the source of light is moving towards or away from the observer (red changes to blue if the motion is towards), and that the orbit of the planet Mercury rotates upon itself to the amount of 43 seconds of an arc per century. Einstein has shown that the orbits of the other planets probably follow suit, though not to such an extent, and that rotating orbits and shifting spectrum-lines can be explained without admitting that the things explained actually happen. These are achievements of the first order, and sufficient to insure the fame of any man. They were the result, however, of standing upon an earth assumed for the moment to be at rest.

How impossible it is to get off the earth, even in imagination, and say anything definite about what one would see from some distant standpoint, is at once evident when one reflects upon what happens to clocks as well as measuring-rods as the result of motion. Clocks move slower when they are carried bodily through space than they do when they are at rest. The general motion makes it more difficult for them to carry an individual motion of their own, such as the turning of their hands about their dials. This

[4] *Ibid.*, pp. 88–89.

is an observed fact, and can be demonstrated by putting a clock upon a revolving turn-table. In scientific language, motion increases the mass, and this retardation takes place whether the time-piece is a piece of machinery or is any other material object whatever.

Therefore, unless the earth be at rest or moving at a known rate in relation to something which is at rest, there is no way of ascertaining the variations that take place in our time-pieces, nor to what extent our measuring-rods either stretch or shrink. Even the length of light-waves will be involved. So it is obvious that under such circumstances we can really measure neither time nor space.

Even if we use that good old clock located in the stomach, we shall still be at the mercy of our motion. If we move in the way of exercise, digestion is speeded up, but if we are carried along bodily it slows up, and it takes us longer to grow hungry. But we will not notice this, for everything about us will also keep slower time, not only our watches but our very thoughts. We won't live so fast. Hence the oft-repeated story of the Relativity Twins, Jim and Joe:

Jim stayed at home and lived fast. Joe boarded a comet, and moved fast—which is the same as living slow. Clearly a case of the more haste the worse speed. So when Joe came back after what Jim thought was a year, Joe had lived only part of a year. Therefore Joe was now younger than his twin.

Einstein can hardly be blamed for our predicament, for shortly after the sun was established as something for the earth to revolve about, scientists began to hint that the sun itself was moving towards the constellation of Hercules. And who shall say that Hercules is not a tripper? One thing is certain. If the earth is moving at all, nobody can at present say how fast or in what direction, and our clocks and measuring-rods are all utterly unreliable bases for celestial speculation.

To what eloquence have we not listened in regard to the magnitude of distant suns and the awfulness of the inter-nebular spaces! We thought we could measure the yard-sticks of Aldebaran, and tell time by the clocks of the Dipper and Orion's belt-buckle. Vain imaginings! Awful these things are, but it is with the awfulness of the un-known. We have not the least idea of the real sizes or dis-tances of the stars, or at what rate the sands of their hour-glasses run. These things cannot be reduced to earthly hours or to terrestrial inches—not if the earth has an undiscoverable motion. Nor can we even guess how fast light actually travels in the sight of the Almighty.

Even on earth, magnitudes may be constantly changing. It may be farther now from the house to the barn than it was yesterday from the house to the moon. And as for time—"Thou hast made my days as it were a span long, and my years are but as nothing in respect to Thee." So you see, these recent discoveries were not altogether unan-ticipated. It may be that the earth does not move forward at all, in which case some of the worst Mistakes of Moses were made by Ingersoll!

Yet Einstein, in theory, wanders continually among the most distant galaxies, promulgating "laws" to be observed by Aries, Taurus, Gemini and Cancer. How does he do it? Very simply. He has a magic carpet known as "the Gaussian System of Coördinates," which he sometimes refers to affectionately as his "Reference-Mollusk." A co-ordinate system is merely a system of measurement, but before examining this particular one it will be well to con-sider another of a less molluskulous variety.

Ordinarily we assume that the earth is a rigid body, so that if we make two marks upon its surface the distance between them will remain practically the same from day to day. We may then draw what we are pleased to call a straight line from one of these marks to the other, and a second line (preferably at a right angle) through the first, making a cross. A third line (which will be up and down) at right angles to the other two, may be represented by a stake driven into the ground at the spot where the other lines cross. Now, if something happens, we can say where it happened simply by measuring its distance from these three lines. Planes will do rather better than lines—say the bottom and two adjacent sides of a cigar-box.

Or we may build a city. If there is a fire there, and we read in the newspaper that it was in the third story of a building at the corner of U Avenue and V Street, we know where the fire was. To tell a stranger where the city is, we reckon from the equator, from a prime meridian, and from the level of the sea, calling these measurements latitude, longitude and altitude. In any case we make use of a coördinate system, and to make it complete we add a clock. Measurements in space are often called "dimensions," or length, breadth and thickness when the dimensions are small. And time used to be called "duration."

Of late, however, mathematicians have taken to calling time the "fourth dimension of space." Of course it is not a "dimension" in the ordinary sense of the word, and those who strive to picture or imagine it as one are simply doing their best to boost the sale of strait-jackets. Time a dimension of space! What joy there must have been in the presence of the angels of Spoof when *that* was put on the air! What pleasure among the cognoscenti as they contemplated the awful havoc that such an expression must make in the credulous modern mind. If this "dimension" had continued to be called "the fourth coördinate," truth would have been served, but where would have been the fun? As it is, a good many of us are a bit confused, like Cowper "boiling his watch and looking at an egg."

Now any coördinate system like those I have described is called "a Cartesian system" if it is not nailed down but may be carried to wherever it is most needed. And then it will be very much like the cigar-box already referred to. But it must remain rigid. This is only possible if the earth is at rest or has a uniform motion in a straight line. If we only pretend that the earth is fixed, or only pretend that its motion is uniform and straight, this rigidity will be pretended also. The actual motion will play havoc with all the times and distances involved. The lines or planes will shrink, or stretch and wriggle and twist. The clock will go wrong. Everything will go wrong. And unless we are absolutely certain about the earth's behavior, we won't know how far wrong, nor in what particular. In other words, our coördinate system will cease to be Cartesian, and become Gaussian. For a Gaussian system is merely one that is totally independable. Gaussian systems have long been used by anglers in telling about the big fish that got away.

Einstein claims that all coördinate systems are Gaussian. He uses no other—at least when in fancy he visits the stars. Ergo, all his stories about stars and space are necessarily fish stories. He calls such a system a "reference-mollusk," because it is like a lot of lines drawn upon the unstable surface of an oyster.[5] The oyster is having a fit. The lines are "imagined," "arbitrary," "curved," and in two sets, one set crossing the other, like the streets of our

[5] *Vide, Relativity,* Ch. XXV, where the lines are imagined as drawn on a warping table-top.

city save that we no longer know where they wander or how long any of the blocks are. To represent the third dimension, you may stick bent pins in the oyster wherever the lines intersect; and for the "time dimension," tie bits of thread to the pins—of no particular length, but with the understanding that the longer a thread is the longer is the time it stands for.

It is now easy to lay down laws for the universe to obey. "We assign to every point of the continuum . . . four numbers . . . coördinates . . . which have not the least direct physical significance, but only serve the purpose of numbering the points of the continuum [that is, the time and space wherein things happen] in a definite but arbitrary manner." [6] In other words, when we measure and describe the universe with the aid of a mollusk, we allege nothing whatever about it save that whatever *is* simply IS. Nature could not help obeying our "laws," for she will be obeying them whatever she does. The mollusk changes wherever you go, for every different place has a different motion—an entirely unknown quantity. And every observer uses his own, individual mollusk, his own measures of time and space, none of which, according to Einstein, have the slightest validity.

You might think it still possible to compare one mollusk with another, but a little reflection will show this to be out of the question. In the first place, you know nothing about your own mollusk except as it seems to you, unless you know how the earth is moving—and it is assumed that you don't. Therefore you do not know how far it is to anywhere, nor how long it takes light to get there, nor how fast it goes. If you want to know at what rate a sun-ray passes the moon you will have either to transport yourself

[6] *Ibid.*, p. 112.

bodily to the moon, or else measure the phenomenon with an earthly mollusk. If you do go to the moon, you can only find out how fast the ray seems to pass, for (as when you were on earth) your assumption that you know if or how you are moving will be but an assumption. Were I now to say that you would find light seeming to pass the moon at a uniform rate no matter from what point of the lunar compass it approaches, I would be assuming that you would find the moon really at rest.

If, as is rather more likely, you remain on earth, you will say that the ray seems to go slower in relation to the moon when moon and ray are moving in the same direction, and faster in the contrary case. But your observation will be warped by your own mollusk, and warped further by your dependence upon light to bring lunar news. The news will be out of date when it reaches you, and you will not know how much out of date, so you can't make the proper corrections. All you can truthfully say is how lunar things would look from the earth if you could see them from the earth the way you think you would see them if you did see them—rather a long shot.

If the earth were really at rest—I don't pretend to say whether it is or not—or moving in a known way in reference to something else which is really at rest, you could correct your own measuring devices, ascertain the true velocity of light, the true magnitudes and distances of the stars, the true force of gravitation throughout the universe, and the actual warpings of mollusks upon moving bodies. If you don't know the state of the earth in regard to motion, you know nothing about speeds and distances but how they appear to you; and even then, if the earth doesn't stand still you cannot account for the behavior of light.

Light is the villain of the piece. If it were only instantaneous we would have something to pin to. It used to be thought that there were such things as "world instants," moments which were simultaneous with each other, a "now" which was Now throughout the universe. Einstein reduces simultaneous events to the simultaneous reception by one consciousness of the news of such events, or else to pure assumptions. From which it is argued that there *is* no "now" which is *Now* everywhere. A thing does not happen until you know about it.

He abandons even the doctrine of the constancy of light's velocity, holding that it is influenced by gravitational fields of unknown strength. When he stretches himself, he destroys not only all possibility of knowledge but all possibility of thought. When bodies are moving in curved paths—and according to Relativity they always are —then, he says,[7] "the idea of a straight line . . . loses its meaning." If so, the idea of a crooked line also loses its meaning, for a crooked line can be known only because of its deviation from the rectilinear. In fact, all ideas of physical forms lose their meanings, once it be admitted that there is no fixture, however unascertainable, and that one reference-body is as good as another by reason of its being good for nothing.

Is there no way out? Of course there is. We must turn from the light of the eye to the light of the mind. In *thought* we have the instantaneous messenger which light proves not to be. We can think of straight lines, because we do. We cannot set our clocks to a universal Now, but we can conceive of such a Now and attribute it to a Universal Consciousness. In such a consciousness alone abides the truth. Without it we are helpless; we would not be. It

[7] *Ibid.*, p. 97.

is folly to look for certainty in physics, for only in metaphysics is it to be found—though it seems unlikely that even physics will long remain in its present confusion, a confusion the boundless extent of which not even scientists themselves appear yet to realize.

But speaking for philosophy, Santayana very wisely remarks: [8] "Relativity does not imply that there is no absolute truth. On the contrary, if there were no absolute truth, all-inclusive and eternal, the desultory views taken from time to time by individuals would themselves be absolute." That is, there would be no standard by which they could be compared and shown to be wrong. Unfortunately, Santayana repudiates his own dictum by holding that there is no all-inclusive and eternal principle separating good from evil. So he elevates the desultory views taken from time to time by society or by individuals into the position of absolutes. Absolute (even in the sense of universal) relativity, either in philosophy or physics, is a round square dyed with pink blue. It is like saying that a relationship exists which is different from any and all relationships. It is unthinkable. And yet—

3. FINITE BUT UNBOUNDED

With a little spizzerinktum, anybody can understand the *theory* of Relativity—which of course does not mean that anybody can understand the Universe—though the matter is somewhat complicated by the fact that there are now at least four distinct brands of Relativity in the market. These may be described as (a) the Common or Garden; (b) the Special or Limited. (c) the General or Einstein; (d) the Goshawful or Bughouse.

[8] *The Realm of Essence,* p. xv.

The Common or Garden sort is used by us all in our daily walk in life. Nobody knows the absolute length of a yard, or how much absolute duration is packed into an hour or a minute. In fact, such expressions as absolute length and absolute duration have no earthly meaning. A yard is the length of a certain actual and material stick kept at a carefully equal temperature (62 degrees Fahrenheit, I believe) in a vault under the supervision of the Bureau of Weights and Measures at Washington, D. C. At least, such is the American yard. In other countries other sticks (they are made of metal) are kept for the same purpose—which is one of comparison. A yard, then, is a length equal to that of an official yardstick. A foot is one third of this, an inch is one thirty-sixth. Or you may chose a footrule for standard, and say that a yard is three times as long and an inch twelve times as short.

Likewise, a pound is a weight equal to a given object, and an hour is an hour as measured by the hands of a specific piece of machinery, called a standard clock. It might be measured by an hour-glass, or by a sundial, but it must be measured by some particular thing. A minute in terms of minutes, is just a minute. If we ask how long it is we can only mean how long in *relation* to some other measured period, arbitrarily selected. If all our time-pieces vary together, nobody will be late for dinner; but to say that they all vary implies a reference to some other time-piece which is held to be going regularly, even though we may know nothing about it. When anything changes its position, we always measure the change from some position which we regard as stable.

The Special Theory of Relativity is more sophisticated. It holds that such "laws of nature" as are good from the standpoint of a given rigid body, are good for any other rigid body which has a uniform motion in relation with the first. Such a motion is called "a uniform translation." An express train going at the rate of sixty miles an hour in relation to the enbankment is said to be in a state of uniform translation in relation to the embankment. It is this Special or Limited theory which is usually described at great length in popular works purporting to "explain Einstein." We have already seen that it (surreptitiously) implies a fixed earth.

The General, or Einstein, Theory of Relativity, does not flatly do away with the Special, but modifies and enlarges it so as to include motion that is not uniform. It dates from about 1915, and concerns itself with the effects of gravitation, which Einstein holds to be due to accelerated motion. If you were in a box being dragged through space otherwise empty, you would feel no inclination to fall towards any side of the box so long as the motion was straight and uniform. You would have—after the thing got started—what you would call momentum, and this would be just sufficient to keep you going along with the box. But if the box increased its speed and kept on increasing it, you would get "heavy" and would fall towards what you would now call the bottom of the box. You would say that you were yielding to the force of "gravity." Einstein would say that you were yielding to your "inertia," or your reluctance to hurry. What you would call your "weight," he would call your "mass"—something which has very little to do with "size," though as mass increases the size of a given body tends to decrease through the packing of the particles together. Newton said that "mass" was constant for a given piece of matter. Einstein says it is no such thing. Mass is the same as "weight" except that weight has to do with that particular gravitational field which relates to the earth, while mass is weight in any gravitational field, that of our accelerating box, for example. If you catch a baseball hot off of the bat, you will notice that it has a new kind of weight. How is this, since though going fast its motion is nevertheless decreasing? Ah, but you are trying to stop it. That is, you are trying to accelerate the decrease, which amounts to setting up an accelerated tendency to move in the contrary direction.

There is another difficulty. In case of accelerated motion such as those cited, inertia drags objects in but one direction. But the gravitation of the earth moves all bodies towards its center, or tries to. A Chinaman falling out of a window in Pekin will fall in a direction nearly opposite to that taken by a Yankee falling out of a window in New York. How, then, can gravitation and inertia be the same thing?

The answer is, "They aren't." Not in every sense. You can't post yourself in any particular spot, and then see the picture as it looks to the parties most interested. To the Yankee, the earth is coming towards him. To the Chinaman, the earth is coming towards *him*. While the Yankee still lingered on the window-sill, the earth (from his point of view) was pushing him up at ever increasing rate. If it had only been content with a given rate and kept hammering at that, his body, once having acquired that rate, would keep it up without effort. But it resists the hurrying, and this resistance manifests itself as weight. When the Yankee fell, the window-sill ceased to push him, and he therefore kept right on with whatever speed he had previously reached. But the earth, coming on faster and faster and faster, overtook him—whereupon his mass suddenly became so great that it hurt. In other words, he was fined for loitering. That is the way it seemed to the Yankee. And that is the way it seemed to the Chinaman in his own case.

The Chinaman's view of the Yankee's case, however, was different. In the first place, it wasn't a view. It was only a report. And in explaining that report to while away the time in hospital, the Chinaman would be inclined to say: "It seems that the Yankee was alle same trying to come to China in a stlaight line. The sill kept him off velly good. Then he leave the sill and come on till the ground plevent."

The Yankee will give a similar account of the doings of the Chinaman. But what will *you* say? It depends altogether upon where you are. If you are in New York, you should agree with the Yankee. If you are in Pekin, you will agree with the Chinaman. If you are off somewhere in space and have eyes good enough to see *both* accidents, you will perhaps hold forth as follows:

"Once upon a time there were a Yankee and a Chinaman, each separated from the other by an earth, and from the earth by a house. Both were trying to get to the center of the earth. Why, I don't know. The houses prevented them. So they stepped out of the houses—out of high windows as it unfortunately happened—and hurried towards the center of the earth, i. e., towards each other. If you ask me why again, I'll scream. They did. Say they had a tendency to accelerated motion, each in his own direction, if you like. I can't be expected to account for the conduct of Chinamen, or of Yankees either. When they got to the surface of the earth, it wouldn't let them go any further. They both stopped. If you don't believe me, I took a motion-picture which proves what I say."

Now I admit that *you* probably wouldn't say anything

of the sort, no matter where you were, because you "know" that in falling from a window a man is pulled to the center of the earth by "gravitation," not by "inertia," and that gravitation is—

Well, what is it? A sort of invisible elastic band bearing the trade mark of Sir Isaac Newton. If you want to be perfectly happy about it you will call it a "tendency." I won't ask you what a tendency is, because you could only say it was "that which" makes a thing do whatever it does do when it happens to do it. Besides, I wasn't supposing that you were you. I was supposing that you were Einstein. The reason we don't all take his view of the foregoing catastrophe is that we are filled with local prejudices. We look at things from the standpoint of New York, for example, and insist that we can at the same time see how things are going on in China. Einstein insists, in theory at least, that a Chinese view is just as reliable as a New York view, or even as a Berlin view. But perhaps you would like to know who is right.

As to that, nobody is right. We are not dealing with motions that purport to be real, but with motions that are relative—relative to the observer. So considered, gravitational phenomena can be reduced to precisely the same "laws," whether they take place in the United States, China, Germany, or in a box hurrying through otherwise empty space. They can all be reduced to the laws of inertia. At least that is how it seems for the moment. The General Theory of Relativity "explains" nothing, whether it be gravity or inertia or the weather or the attractiveness of women. It doesn't even give us a picture of the universe as it presumably is. It merely tells us how we can draw a picture from our own point of view—moving pictures be it understood. Everybody's picture will contradict everybody else's picture, more or less, but the same mathematical equations can be used to describe phenomena in any one of them, and equations are what the theory is intended to produce. Nothing less, nothing more. If you try to make this theory *real,* the whole business becomes subjective. There is nobody left but you.

Bertrand Russell [1] would have it that "the theory of relativity is only the last term, so far, in the elimination of subjective elements from impressions." Nothing could be more misleading, for relativity is distinguished chiefly by the fullness of the recognition that it gives to the part that our subjectivity, our personal bias, plays in our impressions. We are all familiar with the distortions produced by perspective, but it required an Einstein to show us that these distortions apply to impressions of time as well as to impressions of space. In other words, we do not see anything, not even a clock-face or a shadow on the moon, until we see it. If things are in motion, this produces a new set of distortions. We know the sort of corrections that should be made, but not their amount unless we are indeed standing still or know all about our motion. Perspective—there is the General Theory of Relativity in a nutshell.

But if you try to make it a picture of the truth, the whole truth and nothing but the truth, then you adopt the style *d,* or crazy variety. If you are especially eager for your mental undoing, I can recommend the reading of Pavlovsky, who will tell you all about "the innumerable coach" or "ubiquitous motor-bus," which has the happy faculty of moving in the fourth dimension and of being found "at all points along the route at all times of the day," and presumably of the night. And there is Alfred Taylor

[1] *Philosophy,* (New York: W. W. Norton & Co., 1927) p. 155.

Schofield, who describes an interesting folk called "Point Beings," without dimensions, fourth or other, who can see nothing, not even themselves.

Howard Hinton writes a history of Flatland, located on the planet Astria, where there is a right and a left but no up and down. He also writes the biography of Stella, who had to sprinkle her body with flour when she wished it to be visible, because her father had turned all the refraction angles of her substance back to zero. Some prefer P. D. Ouspensky, author of a *Tertium Organum,* intended to complete and supersede the *Organon* of Aristotle and the *Novum Organum* of Bacon. Some of these works were written as "scientific romances," but to be completely devastating they must be taken seriously.

So perhaps it would be best to turn again to Maurice Maeterlinck's *The Life of Space,* which not only describes all the foregoing, but mingles many other wandering notions of the author's own into one highly seasoned cosmic hash the like of which it would be difficult to find elsewhere. It will teach you how to visualize "hypervolumes," "ployhedriods," and "ultra-spiritual entities," to say nothing of "unimaginable monsters, linear, multitriangular and polycubical," at the same time offering enough thoroughly bad science to last a lifetime. Go to Maeterlinck by all means if you would like to think that "mathematics can see further than we can"; that "the higher geometry" is "an extra-human source of information" and introduces us into "a world which is no longer the world of our own creation," where "metageometry" can "discover" that which "is not within us." Maeterlinck was once famous for having created a new sort of thrill of a weird and ghostly sort, very effective on the stage—a thrill of horror. He uses Relativity in order to invoke a thrill more horrible still.

I don't say that he pretends to believe in all these things. He quotes a great deal. And—

But why bother with Maeterlinck? We can revel in style _d_ Relativity without ever taking leave of Einstein himself, who was once asked to describe Relativitiy for a New York daily and replied that it was impossible, "save in the language of mathematics." This was disingenuous, though perhaps he only meant that, translated into the vernacular, it would no longer be safe from criticism at the hands of common sense.

Ought it to be? But in matters of sense we must chose between the _common_ and the _non_. No man has unique faculties, or unique organs of perception. There are only two kinds of reasoning, the good and the bad. Modernist thinkers often lack logic, claiming that it is not logical. They prove it—by the use of logic. What sort of logic I leave you to guess. A man may have exceptional experiences, but when he begins to employ his wits he must be judged by the ordinary rules of sanity, otherwise no argument, not even his own, has any force whatever. He who decries logic does not need to be answered. He admits it himself. Let no one consent to be browbeaten by q-numbers (that is, numbers which are not numerical) and by the square-root of minus one, when, as often happens, they are used to cover a multitude of sins.

Moreover, Einstein often descends to plain German. Thus in his "Message on the Newton Centenary" [2] we find him saying:

"May the spirit of Newton's method give us the power to restore unison between physical reality and the profoundest characteristic of Newton's teaching—strict causality."

[2] _Vide, Nature_, March 26, 1927.

This might be mistaken for a pious belief in the doctrine that effect proceeds from cause. But it is *physical* causality that he was wishing to see rehabilitated—determinism pure and simple, the old mechanist hypothesis. He regrets the escape of human will and aspiration from the prison in which they languished during all the dark ages of the Nineteenth Century, the very prison that he himself has inadvertently done so much to destroy.

Again it was in German, not mathematics, that he described the universe as "finite but unbounded."

"We may imagine," he declares,[3] "an existence in two-dimensional space—flat beings . . . free to move in a plane."

This, of course, is a flat falsehood to begin with. We can "imagine" no such creatures, nor would they be free to move if we could. They could have no thickness, no substance, no material existence whatever. But never mind.

"Let us consider now a second two-dimensional existence, but this time on a spherical surface instead of on a plane. The flat beings with their measuring-rods and other objects fit exactly on this surface, and they are unable to leave it. Their whole universe of observation extends exclusively over the surface of the sphere."

Why are they unable to leave it? Why must their universe of observation extend exclusively over the surface of their sphere, and not beyond it? Well, for one reason, under the circumstances indicated, they would have no powers of observation, and no measuring-rods. Can Herr Einstein "imagine" a rod without thickness? But again never mind. We will be generous, and "imagine" them as having a little thickness, though not enough to attract attention—say an eighth of an inch. They begin to measure

[3] *Relativity,* p. 128 *et seq.*

their domain, and as their rods are nicely curved to lie flat upon the surface, eventually they go clear around the sphere and find themselves where they started from. Their universe has a finite area. And "the great charm resulting from this consideration," says Einstein, "lies in the recognition of the fact that the universe of these beings is finite and yet has no limits."

No limits? But the surface of a sphere *is* a limit, and nothing else. That is its definition. That is what it is for. And yet another limit has been in operation all along to keep these unfortunate Flatties glued to it. Otherwise they might have gone off at a tangent. In effect, while following this charming illustration, we have been "imagining" a second sphere, slightly larger than and enclosing the first— a sphere very like transparent glass. It is this glass and unacknowledged sphere which keeps these Flats with their noses to the ground. They are on the *inside* of this second sphere. No limits indeed! Why, just supposing that they were fleas instead of Flats, and really on the outside. If you think they couldn't hop off you've never been in San Francisco. They might fall back, but they certainly could hop. And if they could only hop just so far it would be because their strength, too, had its limits.

"It follows from what has been said," or at least it follows for Einstein, "that closed spaces without limits are conceivable"; and "as a result of this discussion, a most interesting question arises for astronomers and physicists, and that is whether the universe in which we live is infinite, or whether it is finite in the manner of the spherical universe,"—that is, whether it is finite and yet unbounded.

If ever there was an instance of finite but unbounded nonsense, this is it; and it shows how unbounded are the possibilities of aberration even in the greatest minds when they have an axe to grind. For what Einstein is looking for is not a limited universe for the use of astronomers and physicists—to which he would have been heartily welcome—but a closed system from which God can be excluded—a system which, at the same time, by means of word-juggling, can be said to include the All-There-Is. Not one of the un-geshmatt goyim of our jichness could have beaten this. Evidently Einstein is not a good Jew. And Philo Vance, alias S. S. Van Dine, alias Mr. Willard Huntington Wright, was quite justified when he said, in his inimitable chapter on Mathematics and Murder in *The Bishop Case:*

"Einstein announces that the diameter of space—of *space,* mind you— is 100,000,000 light years, or 700 trillion miles; and considers the calculation abecedarian. When we ask what is beyond this diameter, the answer is: 'There is no beyond, these limitations include everything.' "

The learned Honorary Police Commissioner of Bradley Beach is correct. Einstein in effect announces just that. He would have been quite justified in saying that this particular universe is limited. We find it impossble to get out of it—not because it is unbounded but because it isn't. No physical influence, so far as we can detect, reaches us from beyond a certain (and yet uncertain) weary distance. Our straightest rulers curve and turn upon themselves. No physical manifestation, no ray of light of which we have knowledge, reaches out forever. But to say that the finite is unbounded is to say that the Infinite is limited and to trifle with the lunatic fringes of anti-clerical speech. To add that "space" is curved, indicates a childish desire to confound a proletariat already sufficiently confused. Why violate the ordinary sense of a word, when some other word would so mightily help our understanding? In a detective story, where we wish to be fooled, curved space is a treasure and a gem. Is it possible that we have arrived at a point where we wish to be fooled all the while? Or is this but the bedside manner of pseudo-philosophy?

It were better to change doctors if we wish to get up and about. We will then see that it is not the Infinite but our idea of the Infinite which is limited; and we will try to imagine—not curved space but curved orbits. In fact, the mathematical expression is not curved space, but curved space-time. This is yet more impossible to imagine. But curves of this sort are merely paths. And a curve in space-time is marked only by the successive positions occupied by some moving body, and by the times at which it is supposed to have occupied them.

Mathematicians like to refer to these paths as curves *of* space-time instead of curves *in* space and time. Mathematicians practice a "mystery" in the medieval-guild sense of the word, and have no intention of revealing its secrets to the un-apprenticed. They pretend not to be interested in forces and causes, so they describe the motions that bodies have taken or are expected to take as being "due" to "peculiarities" in the space and time that the bodies encounter or are expected to encounter. It amounts to saying that the path taken and the time it takes to take it are the causes of the taking. The track of a meteor causes the meteor to follow the track—which lies only behind it. My getting up this morning was caused by my getting up.

When a motion has taken place, the mathematician accounts for it by not accounting for it. He says that the moving body merely followed a "wrinkle" in space-time. He remarks to anything whatever: "You're here because you're here, a while ago you were there because you were there, and by-and-by you'll probably be elsewhere for the same reason." Curved space is a curved route. A "curve" in time is a non-uniform motion. We "wrinkle" space only when we want to get rid of the mystery of its disappearance, as in the case of the space intervening between two bodies that approach each other for a reason which we do not feel called upon otherwise to explain. But the only real wrinkles are in the foreheads of those who try to picture themselves as actualities in duration or in the void.

Einstein makes no mistakes in his mathematics. He is not so crude as to put the sign of equality between two non-equivalent expressions. But every problem begins with an "if," a "since," or a "let it be granted." It is with these preliminary suppositions and with subsequent commentaries that he bewilders us, sometimes fools us, and not infrequently fools himself.

In a recent article in a popular magazine it is prophesied that we are going to have less science in the future than in the past. I certainly hope not, for in the past we have had precious little. What we need is not less, but more—less blind faith in vast superstructures of conjecture precariously erected upon an arbitrarily selected group of uncertain facts totally at variance with a host of other facts ignored as inconvenient; and more information, broader foundations, a keener appreciation of our limitations.

Without an absolute (in the sense of standard) assumed for purposes of comparison, Relativity is an empty bag, a method waiting to be applied. Once some fixed standpoint is taken, it rapidly fills with valuable knowledge. When Einstein uses his mollusk experimentally, he treats it as if it were a good old Cartesian cigar-box. Nothing would be easier than to express the fundamentals of his theory in terms consistent with the most orthodox theology. In order to have real motion, we must have a body to refer to which does not move; to have real time we must have Eternity to mark its lapse. This is only another way of saying that reality is that which appears real to the sight of God. It does not mean that we can snatch His sight, or observe the universe from any such commanding position. We have to be content with that which is motionless, changeless, and timeless for us.

Einstein has drawn no picture of the universe. But oh, what a picture he has drawn of man's *knowledge* of the universe—the picture of a knowledge warped with its own incompleteness, as it necessarily must be! He is our greatest secular mind. He has demolished, though somewhat against his will and inclination, the last strongholds of materialism. The danger now is that we shall refuse (theoretically) to believe in the existence of matter at all. Escaped from Charybdis, we are urged towards Scylla. But who urges us? Not Einstein so much as Bertrand Russell. For Einstein, like all great Jews, has an inner vein of true mysticism which, in his best moments, furnishes his own antidote. Thus he has said of Émile Meyerson [4] that he is the only other man living who understands the theory of relativity. And Meyerson holds that all thought is based upon what he calls the "irrationals," which may be interpreted (perhaps a little liberally) as meaning those unthinkables which thinking itself discovers lying outside of and superior to the mind. Bertrand Russell, on the other hand—but he deserves a separate chapter.

[4] Born in Poland about seventy years ago. Worked as chemist under Bunsen in Germany and under Schutzenberger in Paris. He is a naturalized French citizen, and the by no means sufficiently appreciated author of *L'explication dans les sciences* (1921), *Identité et realité* (1912), and *Le déduction relativi*ste (1925).

CHAPTER VI

A VISUAL BLACK DOT OCCURS

I. BERTRAND RUSSELL

SANTAYANA, writing the fourth chapter of *The Winds of Doctrine* in 1913, said of Bertrand Russell:

"It must be remembered that [he] is still young; [that] his thoughts have not assumed their ultimate form. Moreover, he lives in an atmosphere of academic disputation." [1]

Russell was then forty-one, and a lecturer at Trinity College, Cambridge. Still, philosophers ripen slower than do pugilists—or did, before pugilists themselves took to the higher learning—and Santayana was moved to go on, anent *The Problems of Philosophy,* which Russell wrote when a mere babe of thirty-eight: "His book might rather have been called 'The Problems which Moore and I have been Agitating Lately.' Indeed, his philosophy is so little settled as yet that every new article and every fresh conversation revokes some of his former opinions."

Evidently the babe's capacity for devouring reading-matter exceeded his powers of digestion. Nevertheless, Santayana was "charmed." Nobody reading the earlier Russell could help being charmed, though one was "soon made aware that exact thinking and true thinking are not synonymous." Thinking, of course, may be exactly wrong,

[1] P. 111, *et seq.*

or nearly so. But even exactness is something, and Santayana believed that Russell's "inconstancy," his habit of going from one exact thought to another exactly opposite —the habit of going off half-cocked, as it were—was in itself "a sign of sincerity and pure love of truth."

This is very kindly criticism. But it is pleasant to be kind, and to dwell upon those years when "Moore and I were agitating" in that atmosphere of academic disputation so dear to both. The wordy war had been in progress for some time, for Moore—it is G. H. Moore that is meant— was Russell's companion at Trinity long before either was graduated. A perfect pair, these two, like bromium and sulphur, or a tortoise and a hare out to see what was for to see. Few things were likely to escape them both, and if Moore was a little heavy in his discourse, so much the better, since Russell spake only wingèd words.

Good undergraduate talk. Happy, happy times, when days of Damon were followed by Knights of Pythias—a joke which, like the times, should be laughed over seriously, a little sadly, since into its substance has been blown that magic formula, nostalgic with the past, *"O. Henry me fecit!"* Moreover, Christopher Morley once carried it with him when *he* went *Off the Deep End.*

The figure of Russell stands out from the bloodless battle-piece that imagination conjures up, as clear as one of his own exact, inconstant thoughts. He was enthusiastic, "highly verbalized," and versatile. Oh, so versatile! He could (and did) turn from mathematics and moral science (in each of which subjects he eventually won a First) to physical science, philosophy, and politics, all without so much as losing his stride. It began to look as if the world was to have another universal mind, a twentieth-century Leonardo—as if this ship, once it was equipped

~ BERTRAND RUSSELL ~
O, my little Augustine, everything's gone!

with ballast, a rudder, and a pilot, would get somewhere.

As only the base-born are aristocratic, with *la nostalgie de la boue* seldom troubling those who know what it means to go without bath-tubs, he was uncompromisingly republican. The Honorable Bertrand Arthur William Russell—for that is his full name—second son of the late Viscount Amberley, and heir presumptive of the second Earl Russell, could scarcely be anything else in this day and generation. So he worried about "the people" like a true patrician, and in a way that the mere Berts of this world very seldom do. If his second book, published in 1879, was *An Essay on the Foundations of Geometry,* his first, published a full year before, was on the foundations of society—*German Social Democracy.*

In 1917 he was still at it, with his *Principles of Social Reconstruction,* and calling himself an International Socialist in capitals. If 1918 saw the appearance of *Mysticism and Logic,* it also saw *Proposed Roads to Freedom* opened up (verbally, at least) to foot-passengers. He was one of the few outstanding pacifists whose vocal organs were not temporarily paralyzed by the War.

Then, unfortunately for his peace of mind, and after producing a bewildering array of works upon all his other specialities [2] he went to Russia to collect material for his *Practice and Theory of Bolshevism.*

Alack the day! If the theory of Bolshevism and the practice thereof had only been exact and opposite ideas, they might have fraternized in Russell's capacious head.

[2] The list includes *The Philosophy of Leibniz* (1900), *Principles of Mechanics* (1903), *Philosophical Essays* (1910), *Principia Mathematica* in collaboration with Dr. A. N. Whitehead (1910), *Problems of Philosophy* (1911), *Our Knowledge of the External World as a Field for Scientific Method in Philosophy* (1914), *Introduction to Mathematical Philosophy* (1919).

But here was reality. He could by no means stomach it, or get it to behave. China, a year or so later, was decidedly more agreeable—especially old China, which was positively polite. But ever since Confucius even Chinamen have been horribly concrete and practical, and very difficult for academic disputation to deal with. Moreover it was the New China of slant-eyed Sovietism that he had come wishing to believe in; and though this was contrary enough, it wasn't exact even in its etiquette.

The Practice and Theory of Bolshevism, The Problems of China, and *The Prospects of Industrial Civilization* were all bravely committed to paper—the first in 1920, the second in 1922, the third in 1923,—completing the tale of their author's struggles with the cantankerous ways of folks. But the author himself was weary, and the prospects of civilization not so bright as they had been. Decidedly they were disappointing, those trips to the eastward of Eden.

"In China and in Russia," Russell afterwards confided to the readers of an American magazine, "I discovered that I was not as modern as I had supposed."

But we need not here concern ourselves with Russell the sociologist, or with his attempts to poach upon the preserves of Mencken, Shaw, and H. G. Wells. Or with Russell the *pater familias,* who has of late been telling the world how to bring up children behavioristically, while at the same time stoutly denying all the fundamental tenets of Behaviorism. Nor even with Russell the Pacifist, who bravely forced himself back upon Cambridge—this time as a full-fledged professor—notwithstanding his heretical opinion of Mars. As for Russell the mathematician, Russell the moralist, Russell the scientist (he has written with Dora Russell, at least *The A.B.C. of Atoms*), they all are rapidly being engulfed by Russell the philosopher. Reality having floored him, he seems now intent on getting even by showing that reality is really unreal.

In 1913 Santayana found him teaching "that the objects the mind deals with, whether material or ideal, are what and where the mind says they are, and independent of it; that the nature of everything is just what it is, and not to be confused either with its origin or with any opinion about it." [3]

This was pre-War brew, and promised to ripen into rather sound beer. It contained both hops and malt, for its author still believed both in good and in evil.

But in 1926, when Santayana came to write a new Preface for his *Winds,* the epistemological prohibition amendment against belief of any sort had already gone into effect. The nature of things had ceased to be what it is, and had become what it isn't. "What I may have said of the Bertrand Russell of that time (1913)," Santayana was forced to confess, "has little application to the Bertrand Russell of to-day. . . . In the relativity of morals, which I then defended against him, I understand that he now agrees with me. . . . In natural philosophy, too, he is now intent on constructing a system . . . reducing nature to a compound of appearances most of which do not appear. . . . In Mr. Russell's analysis of facts, whether physical or historical, I confess I have little confidence: it is when he derides the existent or plays with the nonexistent that I find him admirable."

So Lenin and Trotsky were not altogether to blame, there was also Santayana with his relativity of morals and his capacity for admiring a tendency to deride the existent and to play with the non-existent. I wonder if it would

[3] *The Winds of Doctrine,* pp. 112–113.

not be still more admirable to deride the non-existent? And certainly it might be well to play something besides ducks and drakes with facts. Anyway, it would seem that between his forty-first and his fifty-fourth year, young Mr. Russell managed in some particulars to go from good to bad, and in some others from bad to worse.

What was the trouble? Why, as I say, Mr. Santayana was the trouble, for one thing. In his *Outline of Philosophy* —published in America in 1917 as *Philosophy, tout court* —Russell frankly acknowledges that it was a gust from the Santayana Winds that helped to blow him loose from his ethical anchorage (the only anchorage he had) and out on that uncharted sea where it is possible "to think that good and bad are derivative from desire." [4] This "desire," I take it, is neither bad nor good. But the question is, has Russell's thought finally become mature enough to make any inquiry concerning it worth while?

Assuredly, yes—mature in the sense of fixed. I never subscribed very heartily to this young-man theory of Santayana's. If, in a way of speaking, Russell is, always has been, and always will be, juvenile, this is only because his is a school that exalts the new, even the raw and green, at the expense of the hoary and wise. Even this may become a stereotyped habit. Personally, he discovered himself long ago—from the moment when a community of mathematical tastes first led him to Leibniz, who taught that each soul was a "monad," an isolated existence with no real means of communication with the neighbors, admitting for the sake of argument that there are neighbors.

Since then his way has been to a certain extent beset with obstacles. Some have delayed him, but none ever

[4] *Vide, Philosophy* (New York: W. W. Norton and Co.), p. 230. All citations refer to this edition of 1927.

turned him aside. As a phenomenon, I think our age has nothing to compare with Russell. Always seeming to follow the latest craze, he had never really followed anything or anybody save his own talent for destruction. Masters he quickly minces into material. In this kitchen no lives are spared. Recently both Einstein and Dr. Watson were permitted to pass the service entrance. When later they appeared in the dining-room it was not as guests, but as hash. Relativity *à la* Russell, Behaviorism *à la* Russell. Everything is *à la* Russell on this menu, with *sauce Anglaise*.

Far from youthful in any hopeful sense, Russell is among the most disillusioned and sophisticated of mortals. Nor is his the sophistication that is weary of baubles, for baubles are all he has ever known. Nor is his disillusionment a freedom from illusion. It is merely a distrust of light.

But I do not wish to be understood as referring to Russell the man, the peaceful neighbor, the good subject of King George, troublesome only in war-time. I am not writing his biography, nor even—save very sketchily—the history of his mind. It does not really matter, for the purposes of this book, how he arrived. The point is, where did he arrive? For I am one of those obsolete persons who believe that philosophy is, or ought to be, something more than self-expression.

For the passions of the mind are cold and pure. The mirror that it holds up to nature should be plain and free from flaws of local origin. Why be content with the wisdom of an isolated individual, when the wisdom of a race, of a species, is available? Wreathe the mirror in garlands if you like, but do not breathe too much upon its surface. Above all, avoid cracking it.

And right here seems at first to lie Russell's supreme distinction. He does try to write philosophy, not poetry or grand opera, and for this endeavor I can never praise him enough. He is so doggedly determined to get to the very bottom of things and to define all terms before using them, that a page of his takes one, for an instant, back to the Thirteenth Century. Santayana even calls this "a new scholasticism"—but with what results to the feelings of the Neo-Scholastics I tremble to think. For something is wrong. Russell is no more a Saint Thomas than Einstein, notwithstanding the etymology of his name, is Saint Peter. Thomas à la Russell is an impossible dish. Russell does not even know the ingredients of the original. If this be Scholasticism, it is Scholasticism sitting out on a limb and sawing off the tree of knowledge. Russell's *Outline,* his *Compleat Angler,* his *Summa Demonologica,* is all out and no line, and might be called *The Reductio ad Absurdum of Everything, or the Shadow which Cast a Man.*

2. THE MALICIOUS DEMON

Russell begins his *Outline* with a chapter on Doubt, and says (page one) that "Philosophy arises from an unusually obstinate attempt to arrive at real knowledge."

One may be too obstinate, I think, and not only look the gift horse in the mouth, but try to pull its teeth. At one time, Science pretended to a worldly, ever-spreading, material infallibility, which was eventually to remove the last trace of a peradventure from human speculations. It accustomed us to demand a type of proof that in fact does not exist—proof that is fool-proof, can be handed out on a platter, and received without the need of wit or judgment. Everything was to be neat, simple, and comprehensible.

So now, though we no longer believe this, some of us continue to go through the motions of looking for certainty of this impossible sort which spares the faculties. All we find, it appears, is doubt, so we determine to make certain of that.

Doubt is a very useful small boy, and should frequently be made to sweep out the store. But it is a totally impossible beginning for philosophy. You cannot be sure even of your doubt if you doubt everything. Until you find yourself in a situation of some sort, already keeping shop, so to speak, you are not called upon to clean up, take an account of stock, or explain how you came to be a merchant. You may be in doubt as to what you are, in doubt as to the nature of your surroundings, in doubt as to the significance and value of life and of the profits possible to the business, but you cannot doubt that you are alive and in *some* surroundings. If you are not certain that things appear to you, and that they appear as they do appear, you are not yet conscious, and philosophical endeavor on your part may be slightly premature. One can hardly say that the inability to think is the beginning of thought.

Not even religion promises a certainty of the automatic, don't-have-to-think species—no, not even the Catholic religion, with its doctrine of Papal Infallibility. The certainties of faith are the certainties of spiritual perception, not of worldly information. Those who believe in Christ and that he has a vicar on earth, must believe that the vicar, when acting as such, speaks the truth. But—aside from the opinions of those who are infallibly certain that Christ has no vicar—I never heard of any doctrine of Lay Infallibility, or any assurance that words, *by whomsoever uttered,* could not be misinterpreted.

At the same time, though we cannot understand the

truth fully and remain human, we cannot deny it utterly and remain rational. Catholic, Protestant, Jew and Hottentot must unite in admitting that Truth exists. For, just as Relativity, to mean anything at all, has to be supplemented by a genuine body of reference, so both Religion and Philosophy presuppose that, in the midst of the chaos of appearance, there is a Reality.

Russell gets off on the wrong foot by insisting—still on page one—that "the first step in defining philosophy is the indication of" its "problems and doubts, which is also the first step in the actual study of philosophy." But obviously the first step is the defining of its certainties. Let us try to compare two opinions, both of which are doubtful. This means that we wish to ascertain which is nearer to the truth. If there be no truth, no opinion could be doubtful. It would, ridiculously enough, become the truth itself. There would be nothing to put it to shame. Truth is something outside of and beyond ourselves, never fully to be ascertained, never to be doubted. We can doubt only the accuracy of ideas, pictures, opinions, designs pretending to represent it.

Russell states the matter much better on page 166, where he says:

"In actual fact, we start by feeling certainty about all sorts of things. . . . Whenever we feel initial certainty, we require an argument to make us doubt, not an argument to make us believe. We may therefore take, as the basis of our belief, any class of primitive certainties which cannot be shown to lead us into error."

It might be well, then, to continue on good terms with actual fact, even when we "begin philosopher." The difference is not a mere quibble. Why suggest that doubt can be the foundation of anything? You cannot begin a profitable search for gold by assuming that there is no such thing as gold. Nor were those very first certainties, which came before we were open to argument or capable of pretense, acquired without reason. Our philosophic career starts as soon as we are born. We already have a broom in our hands when we arrive at that later house-cleaning, the object of which is to get rid of other people's opinions of which we have acquired only the dust. We don't sweep *with* the dust. Nor can we leave go of the broom. Our capacity for doubt is limited. Many books are written for the purpose of removing, or gloating over, doubts which cannot possibly exist. Imagine this "argument" of which Russell speaks, emerging out of a boundless swamp of unbelief! How could we know when we were led "into error" if we had not some way of telling when we were right? Is an argument a certainty more primitive than a primitive certainty, even when it is all composed of doubt? It must be a Bertrand Russell argument. Common arguments start out with certain propositions assumed to be true, or certain data known to be true. They assume the existence of truth and logic. A Bertrand Russell argument assumes only the existence of words.

"The environment causes words, and words directly caused by the environment (if they are statements) are 'true.' . . . How can we get outside [of] words to the facts that make them true or false? Obviously we cannot do this within logic, which is imprisoned in the realm of words." [1]

A word *directly* caused by your surroundings must be one uninterfered with by thought or reflection, a mere reaction, a mere spasm of the vocal apparatus or other group of muscles, like the "ouch!" that follows directly upon the

[1] *Outline,* pp. 262–263.

stick of a pin, or like the squirms that you squirm when tickled in the risibles or elsewhere. As "statement" they all seem to be about on a par. The final arbiter of truth, then, is a squirm or a shriek—possibly of laughter. And yet you can't get at facts with *them*. They confine you to logic. To get out and become logical you must keep still, with your mouth shut. Every time you open your mouth you stick your foot in it. And Russell concludes the "argument" by remarking, "I think the above theory, as it stands, is too crude to be quite true." I heartily concur.

"What passes for knowledge in ordinary life," he says again—we are back once more to page one, and his words, I trust, were caused by his environment, since they are statements, and if caused by anything else cannot be true —"what passes for knowledge in ordinary life, suffers from three defects: it is cocksure, vague, and self-contradictory. The first step towards philosophy consists in becoming aware of these defects . . . in order to substitute an amended kind of knowledge which shall be tentative, precise, and self-consistent." Good! Behold I utter a series of tentative, precise, and self-consistent ouches! But evidently the environment has changed. We are no longer cocksure that we doubt. So possibly we don't doubt.

When a rather more important doubter than Russell— a man named Descartes—ventured to take a first step in philosophy, he began with the famous dictum, "I think, therefore I am." It might have been better to say, "I think, therefore I know that I am," for his thinking was hardly the cause of his being. He did not, I venture to suggest, begin to think before he began to be, and finally think himself into existence. However, he began his thinking with the certainty that he was there and at it.

Aside from this, he went through the motions of doubting everything, even the reality of the external world. It was just possible, he declared, that some malicious demon was furnishing him from moment to moment with a series of illusions—false memories of things that never happened, false experiences of things that did not occur.

I can't see what there would be malicious about such a demon. Accommodating, I should call him. For illusions that could extend and coöperate in this way would cease to be illusions. There can be no such thing as universal illusions. As well speak of universal exceptions. An illusion which answers *all* the purposes of reality, does not differ from reality. It *is* reality. When we say that a counterfeit looks like the genuine article, we mean that it doesn't— not quite. Dealing with a counterfeit dollar does not, in the long run, lead to quite those experiences which come from dealing with a lawful dollar. In one case we go to the bank, in the other to jail. The phony lets us down. It is as real in its way as any dollar, but it promises one set of experiences and leads to quite a different set. If we believe in a lie we are doomed to disappointment. If we don't believe in it, the liar is disappointed. It is because they let us down, or let somebody down, that we call falsehoods false. All errors may be described as prophecies fated not to be fulfilled.

Had Descartes really distrusted his demon and his demon-given memory, he could not have said, "I think, therefore I am." As a matter of fact, what he said was *"Cogito, ergo sum."* Without memory he could never have spelled it, or have fancied that it meant anything. But let us suppose that when he got to *ergo* he had forgotten *cogito,* and that when he wrote *sum* he was no longer aware that he had written *ergo.* False memory, perhaps, stepped in and informed him that he was just finishing the sentence, *Abusus non tollit usum.*

The demon must have been very busy, changing the words already written before Descartes could glance back and discover the mistake. Methinks the demon was taking an unconscionable amount of trouble, and all to no purpose. For after all, a false memory is a real memory, and can only be called false when found to differ from some other memory. A falsehood can't stand alone. Do you think that the demon could have made it true that Descartes, after having written *cogito,* had *not* written *cogito?* Can a demon change the past? To answer such a question you must compare the present with the past. And where are you going to find the past? In some memory, of course. Therefore to speak of a universally false memory is as absurd as to speak of universal counterfeit money.

Memory of any sort is a mystery, for it implies a present that reaches back and includes the past. The most fleeting sort of an illusion implies this, consciousness itself implies this. People look for the immortality of the soul in some far distant future beneath strange skies beyond the grave. But it is immortal here and now, beneath all accidents of *hic et nunc,* all time and change, How could it be aware of time and change were it a part of the flux and not apart from it? The river does not flow past itself, but past its banks.

We speak of the present as the line where future meets the past. It is much more than that. Such a present would have no duration. Nothing could happen in it, and we would be confronted with the enormity of having to say that things happened either in the future or in the past. The real present is more like a narrow rift in the clouds through which eternity peeps out. Psychologists call this felt and experienced present, "the specious present," their reason, no doubt, being the fact that it is the only sort of present that is not specious. It is a conscious moment, utterly immeasurable. Memory is a part of it, or it could not be a conscious moment, could hold no room for consciousness. We cannot think, feel, or imagine a moment apart from all consciousness, for we know of no such moments.

If we say that true memory is one that agrees with deity, we are saying that it agrees with something outside of time and change. Our demon has become God, and the false, through the sheer perfection of its falsity, has become true. There is no such thing as universal error, for error to remain error must be limited. Outside of and beyond it there must be Truth with which it can be compared and found wanting.

Descartes was really wondering if he might not possibly be insane—insane in that complete sense which makes comparison, communication, and all knowledge of others impossible. He was asking himself if he might not be one of those souls enclosed in an impermeable sheath, a windowless house, that Leibniz called monads.

Now we all know that souls may be lonely. But if a soul were lonely to this extent, were there no window whatever, no channel open through which influence of any sort could arrive, then it would be its own creator and sustainer. So Descartes asks in effect if he is not, perhaps, the only pebble on the beach. But this at once destroys the possibility of there being a demon that gives him false memories. It destroys, too, the possibility of his own memories being false. He cannot think falsely because his thinking alone makes the truth. Our philosopher is dabbling in Solipsism. By hypothesis, he becomes God.

I might be moved to permit myself to blaspheme in this

wise, but I cannot possibly permit another to do so, for it implies that *I* do not exist. Not even a God, let alone a Descartes, could create me through a windowless, impermeable envelope. I therefore conclude that if Descartes was entirely isolated in the way suggested he must have been a non-existent Descartes, and that the proper remark for him would have been, "I don't think, therefore I am not."

As a matter of fact, he did not doubt quite so much as he pretended. Nobody as insane as that could retain consciousness. He presumed that he had at least enough true memory to remember his Latin. But it is high time that Bertrand Russell was called back to the stand, for he, too, has objections to offer, albeit objections of quite a different kidney.

" 'I think, therefore I am' won't do as it stands," he says.[2] "What, from his own point of view, he [Descartes] should profess to know is not 'I think,' but 'There is thinking.' He finds doubt going on, and says: There is doubt. Doubt is a form of thought, therefore there is thought. To translate this into 'I think' is to assume a great deal that a previous exercise in scepticism ought to have taught Descartes to call in question. He would say that thoughts imply a thinker. But why should they? Why should not a thinker be simply a certain series of thoughts, connected with each other by causal laws?"

Why not, indeed! Descartes "finds doubt going on." Doubt is thought. Therefore he found thought going on. But he didn't think. Now I do wonder how, in that case, he found out that thought was going on. Evidently he didn't, for he was himself nothing but these very goings on. So it was the goings on which found out the goings

[2] *Philosophy*, p. 163.

on. These goings on were doubts. And a series of doubts, all unfounded, makes bold to discover the series. My own previous exercises in scepticism lead me to doubt it.

But my heart is touched. Think of doubts with no doubter to go to—doubts orphaned, virgin, sweet, standing it may be with reluctant feet where the brook and river meet! I begin to feel awe of this Russell whose "causally connected" unbeliefs can conjure up such a picture.

And what remains of the Cartesian formula? If the "I" in "I think" is objectionable, the "I" in "I am" must be equally objectionable, especially since we learn on page 163 that "when we say 'I think first this and then that,' we ought not to mean that there is a single entity, 'I,' which 'has' two successive thoughts." Evidently "I am" must go. All that is left is "Thinking is going on, therefore thinking is going on." Descartes, however sane and healthy, is out of it. But everything seems to be nice and logical, and tightly confined within the realm of words. It only remains to be seen if thinking is going on as a matter of fact.

For instance, what in the name of dubiety are these "causal laws" which can create a Russell or a Descartes simply by connecting up a series of unthoughts, of doubts *in vacuo?*

"Given certain very general assumptions . . . there are bound to be . . . causal laws. These general assumptions must really replace causality as our basic principles. But, general as they are, they cannot be taken as *a priori,*" that is, cannot be regarded as depending upon something prior to themselves; "they are the generalisation and abstract epitome of the fact that there are causal laws, and this must remain merely an empirical fact," a fact in some one person's experience, "which is rendered probable, though not certain," notwithstanding that it is a fact in experience, "by inductive arguments." [3]

There are bound to be causal laws, then, because we assume that there are such laws. This assumption takes the place of cause in causing things to go. The assumption is not based upon anything prior. It is a generalized and abstract epitome of the fact that there are causal laws, and if this fact is not prior it must be simultaneous or subsequent. So it does not become a fact that there are laws until we assume that there are. Well, after all, these are our laws. We discovered them through experience. Our discovery seems to be what made things obey them. But this is only probable. I should say it was even less than probable. Nevertheless, this improbability is our "basic" principle. An argument, however, soon puts it all right. Once more an argument is more basic than our base. Moreover, this argument is an "inductive" one, that is an argument based upon previously observed facts. The only fact mentioned is the fact that there are causal laws. So we are assured that there are causal laws capable of replacing causality by an argument based upon an observed fact which is uncertain. Anyway, we have assumed them, so they are "bound to be" certain, whether certain or not. And if this isn't a convincing argument, what is? I certainly could not produce a better one—not if I had to prove that causal laws can exist without causality, that very cause, or relation of cause to effect, of which they are the law.

Russell (or rather that certain series of doubts connected with each other by the laws of something which has been supplanted by the abstract epitome of the doubtful

[3] *Ibid.*, p. 150.

certain fact that there are such laws, commonly but dubiously called Russell) thinks it possible that Descartes, notwithstanding the fact that he, too, was a similar series, and no "entity" which could *have* thoughts—thinks that "Descartes knew he was thinking with more certainty than he knew what he was thinking about," and that "this possibility requires that we should ask what he meant by 'thinking.' " [4]

We should ask! There isn't any "I" that can have thoughts, but there is a "we" that can ask questions. But is *is* time that questions were asked.

"Descartes used the word 'thinking' somewhat more widely than we should generally do nowadays. He included all perception, emotion and volition. . . . We may perhaps with advantage concentrate upon perception." [5]

"I think," says Russell,[6] "we ought to admit that Descartes was justified in feeling sure that there was a certain occurrence concerning which doubt was impossible."

I am glad that there is a certainty somewhere. But how in the world do you suppose it happens? I cannot for a moment admit that Descartes knew either that he was thinking or what he was thinking about—not after all we have heard as to the kind of a card he was. Therefore, arguing *a priori,* and yet with due respect to all *a posteriori* considerations founded upon abstract epitomes of absurd generalizations, I am inclined to conclude that the cocksure party was Mr. Thinkerless Thought. He was sure of his own occurrence. Evidently his previous exercises in scepticism left something to be desired.

[4] *Ibid.,* p. 165.
[5] *Ibid.,* p. 165.
[6] *Ibid.,* p. 164.

"Descartes would say," continues Russell, "that when he 'sees the moon,' he is more certain of his visual percept than he is of the outside object."

Marvelous Descartes! He cannot have a thought but he can have a visual percept, the very sort of thought upon which we are concentrating. He is all outside of himself, a mere chain of doubts, yet he is more certain of the inside than of the outside. Moreover—

"This attitude is rational from the standpoint of physics and psychology, because a given occurrence in the brain is capable of having a variety of causes, and where the cause is unusual common sense will be misled. It would be theoretically possible to stimulate the optic nerve artificially in just the way in which light coming from the moon stimulates it; in this case, we should have the same experience as when we 'see the moon,' but should be deceived as to the external source."

My dear Russell! How can you have a source, and a cause, usual or unusual, without causality? And how can you be uncertain of the moon, and certain of a brain and an optic nerve? A brain and an optic nerve are just as external as a green cheese. A moment ago you ruled out a perceiver. Now you are introducing two—one who sees something which he calls the moon, and another who knows that the first perceiver is mistaken, that the moon has been supplanted by a pin, or something scratching at a brain or a nerve. This nerve-and-brain specialist has become an "I." He is the certain Mr. Sure, and the objects of his certainty are pieces of matter. The only piece in doubt is the inconstant moon. It does appear to be a dark night hereabouts. A previous exercise in scepticism should have shown Mr. Russell the absurdity of being absurd. But what is to be expected of one who suggests the possibility of a thought without a thinker? Who unwittingly personifies a thought and gives it all the attributes of a human being, and even of a god? This is not an exercise in scepticism, it is an exercise in self-deception.

3. FEELING THE COLDNESS OF A FROG

My own opinion is that Russell lived in "an atmosphere of academic disputation" a little too long, and that it was academic in the sense of being sophomoric.

"Moore and I" agitate all problems. That is what problems are for. Nothing is stable, not even those premises which we ourselves have laid down. Is it any wonder, then, that we sometimes agitate the premises, to the great bewilderment of freshmen? And so we violate the fundamental law even of honest debate, which reads: "Thou shalt not shift thy ground."

Russell's method is exemplified by the boy who, having first stood on one leg and then on the other, claimed that he had at one time been standing on neither. He devotes half of his *Outline* to describing "Man from Without," in hopes of proving that the man without is within; the second half to "Man from Within," in hopes of proving the man within is without. He talks of motion without anything which moves—his way of expressing a longing for the ineffable. He sometimes speaks as if the world were not only *described by* but *made of* mathematics—like saying that mathematicians are made of their ciphers. We have just heard him speaking of "the standpoint of physics," —of physics, not physicists—and thus bestowing upon a study an ego which he denies to a student, to a Descartes.

"The view which I am advocating is . . . what we call 'neutral monism,'" he tells us, echoing James, on page 282. "It is monism in the sense that it regards the world as composed of only one *kind* of stuff, namely events; but is pluralism in the sense that it admits the existence of a great multiplicity of events, each minimal event being a logically self-sufficient entity." This monism is "neutral" because it takes no sides in the controversy as to whether the stuff is to be called mind or matter. Russell is merely bent upon reducing everything to a simple *one*—alas, not always the same one!

"Pluralism is the view of science and common sense," or at least it is on page 253, where Russell has "no doubt whatever that it is the true view, and that monism is derived from a faulty logic inspired by mysticism." Thus even "neutral monism" meets with the usual fate of neutrals.

But let us take another point of view, as explained on page 300. "From the point of view of human life, it is not important to be able to *create* energy; what is important is to be able to direct energy into this or that channel." This is where free-will goes down for the count—as if anything without force could direct and turn aside another force!

Not even *words* escape the general onslaught. They were "true," you will remember, when "caused by the environment." But they are not necessarily true, it seems, if painted on a sign.

"Suppose you want your hair cut, and as you walk along the street you see a notice, 'Hair-Cutting, First Floor.' It is only by means of induction," says Russell, page 268, "that you can establish that this notice makes it in some degree probable that there is a hair-cutter's establishment on the first floor."

He means that before we do that we must establish the premise, "Where there are barbers' signs there are apt to be barbers." As he is unwilling to admit any principle of constancy in the universe, he labors in vain to make his haircut probable.

"We may go a step further,"—over to page 291. "The past can only be verified indirectly by means of its effects in the future; therefore . . . logical caution . . . should lead us to abstain from asserting that the past really occurred. . . . And since the future, though verifiable if and when it occurs, is as yet unverified, we ought to suspend judgment about the future also." If the past didn't occur, and if the future is not going to occur, one hardly needs a haircut.

"It would be possible," on page 290, "to . . . make matter a structure composed of mental units,"—in which case hair could be all in your eye. "I am not quite sure that this is the wrong view," Russell adds. "Thus matter will be a construction built out of percepts, and our metaphysic will be essentially that of Berkeley."

It will be, save for the trifling circumstance that Berkeley held matter to be a percept in the mind of God.

But Russell, without pausing to distinguish between himself and God, hurries on to remark: "If there are no non-mental events, causal laws will be very odd; for example, a hidden dictaphone may record a conversation although it did not exist at the time, since no one was perceiving it. But although this seems odd, it is not logically impossible. And it must be conceded that it enables us to interpret physics with a smaller amount of dubious inductive and analogical inference than is required if we admit non-mental events."

Odd is no name for it. Physics is even worse off than I imagined if it must admit that non-existent dictaphones may record conversations, or else must admit that it has no physical subject-matter. All that is left is its "stand-point." But—

"In spite of the logical merits of this view," Russell says, "I cannot bring myself to accept it. . . . I find myself constitutionally incapable of believing that the sun would not exist on a day when he was everywhere hidden by clouds, or that the meat in a pie springs into existence at the moment when the pie is opened. I know the logical answer to such objections, and *qua logician* I think the answer a good one. The logical argument, however, does not even tend to show that there are *not* non-mental events; it only tends to show that we have no right to feel sure of their existence."

It's worse than that. If matter be built out of our perceptions, matter becomes indistinguishable from our *knowledge* of matter. Then *everything* is made out of the inside of us. Under such circumstances, to say that we have no right to feel "sure" of the existence of the external is grossly to understate the case. We are left with no reason whatever for even dreaming of the existence of the external—which makes the laws of authorship rather odd. Russell's readers as well as his books are composed of his own mental events—a dreadful drop in sales.

"I have been assuming," he resumes (page 291), "that we admit the existence of other people. . . . Now there is no good reason why we should not carry our logical caution a step further. I cannot verify a theory by means of another man's perceptions, but only by means of my own. Therefore the laws of physics can only be verified by me in so far as they lead to predictions of *my* percepts. If, then, I refuse to admit non-mental events because they are not verifiable, I ought to refuse to admit mental events in every one except myself. . . . Thus I am reduced to what is called 'solipsism,' i. e. the theory that I alone exist. This is a view which is hard to refute, but still harder to believe. I once received a letter from a philosopher who professed to be a solipsist, but was surprised that there were no others!"

It is pleasant to find Russell so cheerful about it, and that he is not (save when acting *qua* logician) entirely duped by his own academic disputations. But think of the mess into which he has led us poor others! We started with logic imprisoned in words, and with words which were caused by the environment and were therefore true. And now there is no environment to cause them, since an environment would be a non-mental event.

"The words 'mind' and 'matter' are used glibly, both by ordinary people and by philosophers, without any adequate attempt at definition," he laments (page 201), and defines them himself by adding: "My own feeling is that there is not a sharp line, but a difference in degree; an oyster is less mental than a man, but not wholly unmental."

This is hardly the question. We don't want to know whether oysters are more or less intelligent than philosophers, but whether either an oyster or a philosopher is an event outside of the mind of the beholder—whether there is any difference between an imaginary philosopher and a real one. Without settling this interesting point, he says, "I think 'mental' is a character, like 'harmonious' or 'discordant,' that cannot belong to a single entity in its own right, but only to a system of entities."

Does he mean that a lone oyster has not a mental character, but that a bed of oysters is a sort of university? But his metaphor inadvertently reflects the truth that matter does not exist in its own right, since it is not harmony which perceives harmony to be harmonious nor discord which detects cacophony in the discordant.

On page 142, however, "the distinction between mind and matter is illusory." I wonder who or what "has" the illusion? "The stuff of the world may be called physical or mental or both or neither, as we please; in fact, the words serve no purpose." Monism remains neutral as between any and all sorts of nonsense. At the same time, Russell will have it that there *is* a distinction between mind and matter, or at least between his own mind and other matters. He is about to try to find some criterion whereby certain events may be said to be more subjective and certain others less. And I may as well say right out that the criterion finally chosen is a footrule. Objectivity is distance measured by matter composed of mental events! Thus:

"Percepts . . . are more subjective in regard to distant objects than in regard to such as are near." (Pages 134–135.) So one *perception* can be nearer and more subjective than another! At the same time, on page 139, "what you see when you see a star is just as internal as what you feel when you feel a headache." Perhaps it was a headache which led Russell to confuse percepts with the objects of perception. Or maybe he is trying to say that our perceptions are less liable to lead us into surprising mistakes if their objects are not too far off. He manages to get this expressed without error on page 132:

"The nearer our starting-point is to the brain"—that is, the nearer the starting-point of the physical stimulus is to the brain—"the more accurate becomes the knowledge displayed in our reaction."

I'll leave you to figure out how one stimulant, or piece

of matter, can be farther off than another if both are strings of your own mental events. But Russell is now in sufficient trouble without being twitted with a mere bagatelle like this. Let us give him back his non-mental events, and set him to measuring the distance of two stars, one of which, say, is Jupiter, and the other the badge of a Chief of Police.

"All measurement is conventional," he begins, "and it would be possible to devise a perfectly serviceable system of measurement according to which a man would be larger than the sun." (Page 300.) So—

"We might, if we were anxious to preserve the word 'mental,' define a 'mental' event as one that can be known with the highest grade of certainty, because, in physical space-time, the event and the knowing of it are contiguous. Thus 'mental' events will be certain of the events that occur in heads that have brains."

The confusion of percepts with the objects of perception was not, then, due to mental confusion, but to a brain-storm. The most mental of all events, the one which is contiguous to its own knowledge of itself, is itself a cranial event! Our ubiquitous event-stuff remains all matter even when it becomes all mind. Its mentality consists in its nearness to matter, to a cranium. Moreover—

"It must be understood that the same reasons that lead to the rejection of substance lead also to the rejection of 'things' and 'persons' as ultimately valid concepts." (Page 243.) So here we have concepts, as well as brains and heads—brains in which events occur and heads which contain (or are said to contain) brains. But these brains and heads are not *things,* they are "events," the very events, no doubt, which occur "in" them. "Everything in the world is composed of events." (Page 276.) And the distance from gray matter which gives to a non-mental event its blessed lack of mentality must be calculated by means of a conventional but perfectly serviceable system of measurements according to which a man may be larger than the sun, and the brain, perhaps, larger still. But, as Dr. Arthur S. Woodward remarks in regard to the brain of Neanderthal man, "We cannot, of course, go by the size."

Russell assures us that "when I speak of an 'event' I do not mean anything out of the way. Seeing a flash of lightning is an event; so is hearing a tire burst, or smelling a rotten egg, or feeling the coldness of a frog." (Page 276.) He also speaks of "the flash of lightning," as well as the seeing of it. But two pages later he goes out of his way to say that "a piece of matter, like a space-time point, is to be constructed out of events." I hope he does not mean that a space-time point, something having neither size nor duration, is a "piece of matter." But perhaps we do not go by size here, either.

What are we to go by? Well, for some little space-time now we have been going by Relativity, the kind of Relativity which Russell advocates throughout his Chapter X, which makes "the notion of a 'place' . . . quite vague. At best," he says (page 109), "you could talk of a given place at a given time; but then it is ambiguous what is a given time, unless you confine yourself to one place. So the notion of 'place' evaporates."

It can hardly be blamed, for Russell, following Maeterlinck, tries to apply relativity of the absolute, unsupported, empty sort. If your cranium is contiguous to such events as these, and you feel inclined to confine yourself to one place,

I shall not attempt to dissuade you. I only hope it does not evaporate. But alas, it does! And with it evaporate all ideas of heads, their brains, their mental events, and such knowledge as is contiguous thereto.

"We are led to construct matter out of systems of events which just happen, and do not happen 'to' matter, or 'to' anything else." (Page 278.)

So a material event is one that doesn't happen. And a non-mental event is an event which, if it happens, must happen next door to an evaporated place which for some strange reason we continue to call a brain. It begins to look as if, after putting all our eggs in one neutrally monistic basket, we had broken both eggs and basket while trying to sort the indistinguishable eggs. Nothing happened "to" any of them, but—Oh, my little Augustine! Everything's gone!

"Matter as it appears to common sense, and as it has until recently appeared in physics, must be given up." (Page 158.) Haven't we been doing it? "We must give up what Whitehead admirably calls the 'pushiness' of matter." (Page 112.) I'm willing, for one. And now, if matter itself can be induced to give it up, everything will be lovely. But at this point there seems to be trouble, for—

"Belief in the unreality of matter is likely to lead to an untimely death, and that, perhaps, is the reason why this belief is so rare." (Page 168.) Reason enough. So, after all—

"We cannot dismiss the common-sense outlook as simply silly, since it succeeds in daily life; if we are going to reject it in part, we must be sure that we do so in favor of something equally tough as a means of coping with practical problems." As for instance, what?

"From the standpoint of pure physics, matter is only an abstract mathematical characteristic of events in empty space." (Page 146.)

This is tough enough, in all conscience. I shall try to avoid events in "empty" space. But I no longer see how those who refused to believe in the reality of matter met with their untimely deaths. Do you mean to tell us, Mr. Russell, that they were run over by the abstract mathematical characteristics of events in empty space? Go to!

A philosophy which described "things" as "events," meaning real events, might have some point, even a *point d'appui*. Aquinas himself says [1] that "an effect shows the power of cause only by reason of the action which proceeds from the power." But matter, deprived of its pushiness, gives up the ghost, and fades away into the awful No-Ness of the Not. Nevertheless—

"One sometimes, under the influence of indigestion or fatigue, sees" at least "little black dots floating in the air. In such circumstances you would say 'I see a black dot,' but not 'there is a black dot.' " But Russell insists that "a moment's reflection shows that both 'I' and 'see' take us beyond what the momentary event reveals." (Pages 207–208.) So he removes the "I," as in the case of Descartes, and substitutes "visual" for "see." He avoids saying that anything "exists," but suggests "occurs,"—thus hinting that these dots of dyspepsia can occur without occurring to the dyspeptic, and thus manage to exist without existing. Nobody sees it, but yet a visual black dot occurs.

"Can we say, on the basis of immediate experience, not only 'a visual black dot occurs,' but also 'a visual black dot is cognized?' My feeling," says Russell, "is that we cannot. When we say that it is cognized, we seem to me to

[1] *Contra Gentiles,* iii, 2 xix.

mean that it is a part of an experience." (Page 208.)

Precisely the way it seems to me. But, if it is not a part of his experience, can a dyspeptic, upon the basis of immediate experience, say that a visual black dot is occurring? I wonder how a dot can be visual, black, or even dotty, unless it is seen. No matter. For—

"It is not from the logician that awe before truth is to be expected. . . . Just as the grave-diggers in *Hamlet* become familiar with skulls, so logicians become familiar with truth." (Page 254). I see. Familiarity breeds contempt, and "the hand of little employment hath the daintier sense." Logicians, as Russell understands them, are morticians. Yet they meet with Truth only to bury her—alive. Anyway (same page), "the question of truth and falsehood has been wrapped in unnecessary mystery." Off with the wrappings, then!

"If there be a world which is not physical, or not in space-time, it may have a structure which we can never hope to express or to know." (Page 265.) Never completely, no. Here is Truth, naked at last. How unmysterious she is!

As a matter of fact, the critic of Russell is not called upon to explain a philosophy, but to diagnose a disease— a disease of the times. We are living at the end of a long series of "reformations," and this last one has gone wild. It has credophobia. Undoubtedly it will recover. The determination to defy all reason, all authority, all tradition, all standards, and to believe nothing which anybody ever believed before, runs its course like any other epidemic. At the same time it must be admitted that Bertrand Russell is a very sick man.

CHAPTER VII

THE WINNOWING FAN

I. JOHN DEWEY

CALL in Dr. Dewey. And what does he prescribe? Nobody seems to know. Last fall his seventieth birthday was celebrated with considerable pomp and ceremony in New York. Speakers praised his educational work, as was but just. But nobody attempted to explain the philosophy that he offers to adults.

A world-famous professor said recently in a private conversation: "I have been making an intensive study of John Dewey for the last six months, trying to find out what he means. I give it up."

This cheered me wonderfully when I heard of it. And now I wish to cheer the professor in return. So I advise him to open Eddington's *The Nature of the Physical World,* at page 221, and read:

"When we encounter unexpected obstacles in finding out something which we wish to know, there are two possible courses to take. It may be that the right course is to treat the obstacle as a spur to further efforts; but there is a second possibility—that we have been trying to find something which does not exist."

This second possibility should not be lost sight of in approaching Dewey. Undoubtedly he has an object in view when he writes. He wishes to explain everything in

— John Dewey —
He should have remained old-fashioned

the heavens above, in the earth beneath, and in the waters under the earth. The inexplicable, then, must be done away with. Naturally the explicable goes with it. Here is another attempt to write a Preface to Genesis. But it turns out to be only an account of a non-existent universe. If his fundamental tenets, in so far as they may be guessed, were true, there wouldn't be any universe. To search his most characteristic pages for meanings which mean anything—it is often but lost labor that ye rise up early and go so late to take rest.

That he has given headaches even to his philosophical confrères, he himself admits. In *Experience and Nature* [1] he laments that there is a widespread tendency to deny that he is what he thinks he is. "The empirical method employed in this volume," he declares, "has been taken by critics to be simply a re-statement of a purely subjective philosophy, although in fact it is wholly contrary to such a philosophy."

I do not wish to be pedantic, or to lose the spirit in following the letter. But with a man of Dewey's eminence one really has a right to suppose him aware of the importance of words, and a philosophy "wholly" contrary to a "purely" subjective philosophy must be a wholly and purely objective philosophy. This is a wholly and purely impossible philosophy, one which supposes everything to be outside of us, and that we can study it without using anything inside of us. Nor does Dewey help matters when he calls this an "empirical" method, since the word "empirical" signifies, "founded upon experience." Is Dewey's experience, then, wholly and purely outside of himself?

He goes on to call his philosophy "the denotative

[1] New York: W. W. Norton and Co., edition of 1929, note to page 16.

method," or "empirical naturalism," "naturalistic empiricism," "immediate empiricism," or "naturalistic human-
ism," all of which are nice names, and seem to indicate
that here at last we have a real philosopher, who, while
admitting his own existence, admits also the existence of
nature, and listens attentively to what the wild waves are
saying. But have we? He would say, "Yes." But there is
a catch in it. He listened too long to what William James
was saying. These magic formulas, which I have enclosed
in quotation marks, are Dewey's "Open, Sesame!" Yet
they open nothing. Their magic has been extracted. But
before trying to discover how it was done, let us go back
a little to things more entertaining.

John Dewey was born at Burlington, Vermont, in 1859
—a Calvin Coolidge type of person. His A.B. was con-
ferred by the University of Vermont in 1879, his Ph.D.
by the Johns Hopkins, in 1884. He was elevated to the
Chair of Philosophy at Columbia in 1904. And in 1929
he secured the great plum of the English-speaking modern-
istic academic world, the honor of delivering the season's
Gifford Lectures, at Glasgow.[2]

[2] The style and purport of these lectures may be seen by the
following sample: "The mind is within the world as a part of the
latter's own ongoing process. It is marked off as mind by the fact
that, wherever it is found, changes take place in a directed way, so
that a movement in a definite one-way sense, from the doubtful
and confused to the clear, resolved, and settled, takes place."
No doubt changes do take place in a directed way wherever mind
is found. But to say that mind is within the world and is a part of
the world's own ongoing process is to say that it is produced by
the world, that the doubtful and confused give birth, without any
help, to the clear, resolved and settled. Thought of this sort hardly
contributes to any such movement. It comes from a mind which is
itself doubtful and confused. Doubt and confusion are not in the
world, but in man's idea of the world. Even so, "process" could have
hardly brought the world into existence, let alone bringing the clear
and resolved into existence, if its force lay merely in its own doubt
and confusion. The only truth in the passage lies in the fact that

Once he was much interested in Darwinism, but it is in his theories on education, and in his practical endeavors to "rear the tender thought, and teach the young idea how to shoot," that his real superiority has been manifested. Strangely enough, though called upon to act as architect of the new school systems in Soviet Russia and in China, as a pedagogue he is something of a medievalist. His insistence upon craftsmanship, upon the education of the *hands,* certainly smacks pleasantly of the ancient guilds. Here, in fact, is one whose heart, whose best, has gone out to children, leaving only the husk for us tedious grownups. He teaches how to shoot much better than how to

the operations of human mind may be discovered by discovering objects and institutions shaped in conformity with human purpose. That is, the effects of the human mind are a part of the totality of effects to be noted in the world.

But Dewey says that "from knowing as an outside beholding to knowing as an active participant in the drama of an onmoving world, is the historical transition whose record we have been following." We are thus asked to believe that a mind that was "within the world" and a part of the world's "own ongoing process," was guilty in its youth of an "outside beholding"! So we are compelled to drop the first part of the paragraph before arriving at the end. Even then it is far from making good sense. The mind is undoubtedly outside of the flux when it beholds. But it does not progress from this outside beholding, this aloof contemplation, to a later more active participation in the drama of life. The history not only of individuals but of races and civilizations has progressed in a manner precisely opposite to this. We are more active in childhood, and more thoughtful in age. Evidently Dewey had vaguely in mind the progress from a deadly sort of metaphysics, to animalism. But this is hardly to be called a progress from confusion to clearness, nor is it so rejoiced over, save as an indication that we have passed through the worst of the crisis and may be expected to mend. It is society's imitation of the senile attempt to plunge into action sometimes made by frightened old age, the dizzy grasping at salvation by carnival which marks the degenerate end of civilizations—perhaps even our own—as they make way for their successors. But it is a very unlovely moment. The human being does not make a nice animal. Nobody wants to live in a neighborhood of this sort. It would soon be fatal even to the human animals themselves.

think. All must hail him, however, as a sincere and honest man, whose mental befuddlement, at times distressing, is genuine.

Even his educational theories are expressed in lamentable language, which paws after rather than grasps the idea. His own peculiar flavor is a very remarkable tastelessness—I mean a lack of raciness, of savor—a flatness like that of unleavened bread. The color of his aura is drab. But let no one despise drab. It is from drab flint that some of the brightest sparks are sometimes struck. Only the jaded palate—to return to the original metaphor —demands incessant spicing. Dewey might not have made a good President, though he is politically rather than philosophically minded, but he would have made a splendid Calvinist. I should like to have seen him in Salem Witchcraft days, hobnobbing with Cotton Mather. Puritans ought to remain Puritan, if only for the sake of the picturesque.

But Dewey was fated to be born in the Jimsian Epoch, and to survive into the Post-Jimsian—if we really are yet *post,* which I sometimes doubt. Under the circumstances, a flint is bound to become sicklied o'er with the very pale cast of an alien earth. It cannot take the gorgeous dyes of decay. The more it is dipped the duller it becomes. It is only as a person (happily undyed), and not as a prophet, that Dewey retains his color.

He, who should have remained old-fashioned, can even write: "I have not striven . . . for a reconciliation between the new and the old. I think such endeavors are likely to give rise to casualties to good faith and candor. But in employing, as one must do, a body of old beliefs and ideas to apprehend and understand the new, I have also kept in mind the modifications and transformations that are exacted of those old beliefs." [3]

Though compelled to approach the new burdened with the old, he has kept in mind the modifications and transformations that are to be exacted of the old. He doesn't bother to ask himself if it might not, now and then, be the new which needs to do a little sidestepping. Love is truest when it's newest; and so, apparently, are philosophy and science. This is an interesting credo, though it makes a heretic of Wall Street, whose predilection for well-seasoned securities is notorious. By "old" Dewey evidently means "Victorian," or "Newtonian" at the furthest. His mind seldom runs back to vivid images of a more remote past.

"I believe," he goes on in this same page, "that the method of empirical naturalism . . . provides the way, and the only way . . . by which one can freely accept the standpoint and conclusions of modern science."

Coming after Einstein, this touching faith in the permanence of the ephemeral, this dogged and vain determination to keep up with the times, is weird. *Modern* science has arrived at no conclusions. So it is a conclusion to have no conclusions, the standpoint of having no standpoint, which "empirical naturalism" accepts.

In 1926, when Eddington was in the act of preparing one of his Gifford lectures for the following year, he wrote: "The new Quantum Theory [relating to the structure of atoms] originated . . . in the autumn of 1925. I am writing . . . just twelve months after. . . . That does not give long for development; nevertheless the theory has already gone through three distinct phases.

[3] *Experience and Nature,* p. ii.

. . . My chief anxiety at the moment is lest another phase of reinterpretation should be reached before the lecture can be delivered." [4]

His apprehensions were not fulfilled. It was *several months* after the lecture that the theory entered upon a fourth phase—and Bertrand Russell says that the law of the quantum is the only law of nature that has ever really been "discovered"! But listen to Dewey:

"Empirical naturalism . . . provides the way, and the only way . . . by which we can be genuinely naturalistic and yet maintain cherished values, provided they are critically clarified and reinforced. The naturalistic method . . . destroys many things once cherished; but it destroys them by revealing their inconsistency with the nature of things—a flaw that always attended them and deprived them of efficacy for aught save emotional consolation. But its main purport is not destructive; empirical naturalism is rather a winnowing fan. Only the chaff goes, though perhaps the chaff had once been treasured." [5]

This sounds rather plausible, as if it were merely a long way of saying "We profit by experience." But it is all wrong. We do not profit by experience if we accept every "conclusion" as it arrives, or even allow it to act as judge. Experience shows that most new things are but fads. It wasn't so very long ago that physicians thought that salivation was salvation. The list of such passing errors is endless. The "nature of things" with which they were consistent was limited to that kink in *human* nature which leads man to make a new kind of ass of himself whenever possible. The present cannot and does not act as a "winnowing fan" upon the past. It is always

[4] *The Nature of the Physical World*, pp. 206–207.
[5] *Experience and Nature*, pp. ii–iii.

called upon to meet the blast of that winnowing fan, which is in the hands of the past. A very, very little of it usually does, and so finds its way into the past's treasure-house. Are our "cherished values" to be finally appraised, and maybe—ah, the rare chance!—reinforced by conclusions held to be conclusive for sometimes as long as twelve months?

I am afraid that some of the values that Dewey is trying to cherish are on the wrong side of the fan. And I hope that Heisenberg, who is the party who first hurled "quantum phenomena" into the arena, enjoyed the emotional consolation of his theory during the few weeks that elapsed before the other boys (their names are Born, Jordan, Dirac and Schrödinger, to mention only the ringleaders) began to send it through its other phases. Not that I mean to belittle the work of these great mathematicians, or to imply that the existing body of knowledge should not expand to accommodate all new experiences. If Empirical Naturalism aims to keep on accepting science's theories as to the value and meaning of these experiences, however, it ought not to publish books. What it needs is a ticker.

Dewey's own emotional consolations are bound up chiefly with the values current in his youth—the world of Darwin, et al. It was then that he acquired the habit of thinking that he was scientific and up to date. It was then that he started winnowing away the beliefs of other people. He has tried to keep his fan moving forward as well as around. But it has become a chimerical fan, which once blew most of the solid ground from under its own stand, and now belongs to that part of the past which has all but blown away beneath the erosion of ancient and eternal winds.

What I find even more pathetic, is Dewey's frequent description of science as an "art." For this idea makes philosophy, which is only science with a larger subject-matter, an art also. And whatever else Dewey may be, he is certainly not an artist. Should he be one?

Of course he should, if by the art of philosophy you mean the art of thinking. Then why not say dialectics, which is its proper name? We have already seen what an overworked vocable "art" has become—so much so that as used in connection with philosophy it is beginning to have all the characteristics of an ancient wheeze compelled to do service as a poor excuse.

The word "art" carries with it the suggestion of something preëminently individual, subjective; of something made up, appealing to the heart rather than to the mind, voicing the citizen's eternal protest against society, his lament for liberties necessarily (and frequently unnecessarily) lost—his residuary unwillingness to *be* a citizen.

No doubt the theories, whether of philosophy or science, are also made up. Even observations contain their need of the subjective. No doubt, either, that art is to some degree objective when it has a public. Nevertheless, there is a difference in emphasis. For it is the glory of the artist to color his work with himself, trusting to others similarly colored to understand him. But it is the glory of the philosopher and the scientist to work in black and white, eliminating as well as they can those peculiarities due to one particular person, time, and place. Philosophy and science are therefore (in their highest manifestations) collective products. And if you insist that art in its highest (and more especially in its classical and its folk-song and fable) manifestations is also collective, the work of whole periods and races, I still can answer:

"Nevertheless, the goals are different. Or if not the goals, then the routes. Art always expresses the fact that we are lonely. Science and philosophy should show us that we nevertheless have neighbors. And though art does the same, even pleading that our very loneliness gives us something in common, it does not point to the similarity of our perceptions, but to our feelings."

That is the fundamental difference. But it is sadly complicated by the many meanings which are given to words.

Thus art usually concerns itself to a certain extent with the sensuous. It sometimes soars above this, and even identifies itself with religion. It may deal with emotions coming from the contemplation of pure beauty of the most spiritual sort. But "beauty is truth, and truth beauty" only in the final synthesis. How different they are upon the lower levels, the intermediate stages of the pilgrimage! Though we thrill with thoughts and can think of thrills, thoughts and thrills are not quite the same things.

Art, science, philosophy, are all rebels when born. They are then alike in partaking of the new. But they never unite again until all have found their freedom in the All Which Is. And this they never quite reach. It is this likeness at birth which makes them so easily mistaken for each other—like the three children of one couch, who are scarcely distinguishable as babes, show but little family resemblance in the middle years, but draw near to each other once more in age. It is well for the old philosopher to grow melodious, and for the old poet to grow philosophical. But it is seldom well for either quite to forget his own proper métier. When we say that all that hath life hath art, we only mean that all that hath life hath birth.

But my chief objection to philosophers who call themselves artists is that they flatter themselves. If they think they are artists as poets, painters, or even novelists are, let them read a few good novels and poems, and visit a gallery or two. With the possible exception of Santayana, Schopenhauer and Nietzsche, which of the moderns could make good their boast? And even if all could make good, it would yet remain true that a literary artist can, without exciting much remark, turn out a most wretched system of thought.

"The failure to recognize that knowledge is a product of art accounts for an otherwise inexplicable fact: that science lies to-day like an incubus upon such a wide area of beliefs and aspirations," says Dewey." [6]

Evidently this bit was not produced by the art of dialectic, for a moment ago we were to "accept" science's "conclusions," and now they are become an "incubus." I am afraid that this science has turned artist, since it is producing nightmares. One thinks of the *Murders in the Rue Morgue*. But of course a conclusion becomes less of an incubus if you don't believe it's so. Dewey uses "conclusion" artistically.

"If modern tendencies are justified in putting art and creation first," the tale continues, "then the implications of this position should be avowed and carried through. It would then be seen that science is an art, that art is practice, and that the only distinction worth drawing is not between practice and theory, but between those modes of practice that are not intelligent, not inherently and

[6] *Experience and Nature*, p. 382. This is Dewey's chief philosophical work, and the only one to which I shall refer. It was first published in 1925, but all quotations are from the edition of 1929.

immediately enjoyable, and those which are full of enjoyed meanings. When this perception dawns, it will be a commonplace that art—the mode of activity that is charged with meanings capable of being immediately enjoyed—is the complete culmination of nature, and that 'science' is properly a handmaiden that conducts natural events to this happy issue." [7]

Science is the handmaiden that conducts natural events to the happy issue where they can be enjoyed. But we have just learned that science is an art. It is, then, itself a mode of activity charged with meanings capable of being immediately enjoyed, which therefore need no conducting. I wish that the same could be said for Mr. Dewey's exposition. But I see that we go from joy to joy when engaged in intelligent activity. It is a fine idea. Does it, however, justify one in saying that the distinction between practice and theory is not worth while? One would think that all distinctions that are not distinctions without differences were worth while. I would even distinguish between making art, and being fed with it; between composing or playing a sonata, and listening to a piano-recital.

It is possible to dig considerable good sense out of Dewey's words right here, but I do not find the digging immediately enjoyable to any noticeable extent. Perhaps my activity is not sufficiently intelligent.

Let us by all means put creation first. Thoughts come in flashes, whether they be the themes of symphonies or the solutions of mathematical puzzles. Practice makes perfect, and with practice comes discipline, skill, art. But it is by no mistake that we call art in this sense "painstaking." Are we not to distinguish between the pleasant

[7] *Ibid.,* pp. 357–358.

gusto that accompanies the birth of a notion, and the hard grind of working it out, which one endures, sustained only by the pleasure of the hope of a future reward? Amateurs seem to find this work a great and sufficient pleasure, but I notice that real artists continually grumble about it. Let it be granted that art, science, and philosophy all begin with inspiration. But this inspiration, taken alone, is unworthy to be called either art, philosophy or science. They all demand "art" in the sense of "skill," of course.

But I think that Dewey's way of putting it is misleading, that it leaves the impression of a justification for slovenly thinking and lazy doing. His lack of distinctions appears to have done nothing but increase the mystification. But as the error, if any, is again chiefly in emphasis, and in a looseness of language which itself seeks to give a vague enjoyment, as of art, by tempting us not to bother with thought, let us pass on to where the same habits of mind lead to troubles of a much more serious and fundamental sort, as when he says:

"Thus [in the disappearance of distinctions] would disappear the separations that trouble present thinking: division of everything into nature *and* experience, of experience into practice *and* theory, art *and* science; of art into useful *and* fine, menial *and* free."

I flatly refuse not to distinguish between the menial *and* free, whether in the practice of art or of husbandry, especially since here is that very difference between intelligent and unintelligent activity which we have just been invited to note. And not to distinguish between practice and theory is to confuse concepts and percepts, self and environment. But to say that the trouble with present-day thinking comes from the distinctions it makes, is undoubtedly a mode of activity that is charged with meanings capable of being immediately enjoyed, so it must be art—the art, I should say, of unconscious humor. But it is not the truth. For more than a quarter of a century now, modern thought has been engaged chiefly in the (I should say unintelligent, and therefore neither artistic, nor scientific, nor philosophical) activity of abandoning distinctions everywhere. And if any remain, they too, are about to go. We draw nigh once more to the dismal swamp of Everythingisthesame—in the hope, perhaps, that here trouble will not be able to follow us.

2. A HOUSE TO LET

According to Blake, matter was created by Urizen, and it must be admitted that if your reason is good for anything it will at least be able to discover that matter was created by something. But Blake, who demonstrated the unity of art and philosophy by being one of the world's great artists and one of the world's worst thinkers, had a remedy. He named it Los, old Sol spelled by a looking-glass, the inner flame, almost your soul if you had one. These two "gods" had but to come together, and matter would disappear, for then you would have Los-Urizen. What more could you ask? [1]

Dewey goes Blake one better, for he would have it that this loss, this union, has already taken place.

"We begin," he says, "by noting that 'experience' is what James called a double-barrelled word. . . . It is 'double-barrelled' in that it recognizes in its primary integrity no division between act and material, subject and object." [2]

[1] I am indebted to a suggestive article on Blake by Cortlandt van Winkle in *The Commonwealth* for October 23, 1929, p. 649.
[2] *Experience and Nature,* p. 8.

So experience in its primary integrity is double-barreled because it is single-barreled. Nor can it be argued that this "primary integrity" refers to some entity back of creation, for it is the "primary integrity" of "experience" we are supposed to be dealing with, and obviously an entity standing alone precludes experience. If it has experience it ceases to be alone, and its primary integrity is shattered. Dewey even proceeds to shatter the primary integrity of his own absurdity by adding:

" 'Thing' and 'thought' as James says in the same connection, are single-barrelled; they refer to products discriminated by reflection out of primary experience."

From which I gather that reflection discovers that thoughts are not things. It puts them in different barrels. So it was the lack of reflection which gave that primary integrity to experience, and could see no difference between subject and object, act and material, a jump and Mark Twain's jumping frog. I shouldn't wonder.

But Dewey is unwilling to trust to reflection—perhaps because it really has a sort of primary integrity, though not so primary that the self whose act it is can be called self-subsistent in the midst of its own uncreated, unexperienced world. "Empirical method"—i. e., his own method—"alone takes this integrated unit [of primary, unreflected-upon experience] as the starting point for philosophic thought." In other words, Dewey's philosophic thought starts by being unreflective, in the single-barreled double-barreled stage, when you can't tell the difference between yourself and a shotgun. "Other methods begin with results of a reflection that has already torn in two the subject-matter experienced and the operations and stages of experiencing. The problem is then to get together again what has been sundered—which is as if the king's men started with the fragments of the egg and tried to construct the whole egg out of them." [3]

This is delicious, as I hope the egg was. Non-Dewey methods begin with Humpty Dumpty busted. To them there is a difference between the white and the yolk, and even between eggs, walls, falls, king's horses and king's men. How shall we recover from the effects of this devastating reflection? We shan't. Perhaps we don't even want to. But Dewey does.

"For empirical method the problem is nothing so impossible of solution. Its problem is to note how and why the whole is distinguished into subject and object, nature and mental operations. Having done this, it is in a position to see *to what effect* the distinction is made: how the distinguished factors [such as eggs, men and horses] function in the further control and enrichment of the subject-matters of crude but total experience. Non-empirical method starts with a reflective product as if it were primary, as if it were the originally 'given.'" [4]

Exactly! You see, Mr. Dewey, that what you call the product of reflection *is* the originally given. If ourselves and something else, subject and object, experiences and something to experience them with, had not been given, we should have Los-Urizen, we should not have had any, or anything else, to loose. These distinctions were not created by reflection, but discovered by reflection, beginning with that first glimmer of reflection which made us aware that we were alive. You are simply playing Blake, and trying to make Reflection more primary than primary experience—so primary, in fact, that it can first distinguish distinctions which you imply do not exist, and

[3] *Ibid.,* p. 9.
[4] *Loc. cit.*

can then enrich the crude total by distinguishing the factors and functions of these distinguished, indistinguishable, distinct distinctions. The fallen egg is getting pretty high.

It is going to get higher, for Dewey believes that "the chief obstacle to a more effective criticism of current values lies in the traditional separation of nature and experience, which it is the purpose of this volume [*Experience and Nature*] to replace by the idea of continuity." [5]

We are now at the very core of Dewey's difficulty. He wants to feel that "experience is not a veil that shuts man off from nature" [6]—a most laudable ambition. But he is afraid that if he admits that nature and experience (i. e. matter and mind) are really separate, he can find no way for any communication between the two.

"To non-empirical method . . . object and subject, mind and matter (or whatever words and ideas are used) are separate and independent. Therefore it [non-empirical method] has upon its hands the problem of how it is possible to know at all; how an outer world can affect an inner mind; how the acts of mind can reach out and lay hold of objects defined in antithesis to them. Naturally it is at a loss for an answer, since its premises make the fact of knowledge both unnatural and unempirical." [7]

Well, the fact of knowledge *is* both unnatural and unempirical, or at least supernatural and supra-empirical. If there were nothing but matter, the fact of knowledge would never have become a fact. Nor would a human mind alone have been able to know anything, or furnish itself with anything to know. If a critic is called upon to

[5] End of the preface, dated 1929.
[6] *Ibid.,* p. iii.
[7] *Ibid.,* pp. 9–10.

show how man could have created himself, his Creator, his environment, and his knowledge, I resign.

I also admit that a "non-empirical" method, asked how an outer world can affect an inner mind, would be "at a loss for an answer." But then there isn't any such a thing as a "non-empirical method," any more than there is such a thing as an "empirical method." Every philosopher, every cook and bottle-washer, must take note not only of his own individual, "empirical" adventures in this world of masterpieces, but of the fact that there is something "non-empirical" astir about him. A purely subjective or a purely objective method is impossible.

But what worries Dewey—and has worried many another before him—is the conviction that there is something like a contradiction in terms in saying that matter affects mind, or mind affects matter. If mind and matter have nothing in common, they ask, how can there be any traffic between them? And the answer is, there can't. But this stumbling-block is merely a hold-over from an out-moded physics, which taught that matter was something absolute and self-existing. Once admit that beyond matter there is something to which we are akin, that it is with this that we are in communication—that matter is indeed a veil, though not quite concealing that over which it is thrown—and the supposed contradiction in the common-sense view at once is eliminated. If these modernists were only modern, I should hasten to join up with them. But they are neither modern, nor medieval, nor ancient. They insist upon falling between all stools which have ever been set up, and collecting only such rubbish as has been abandoned to the floor—to call bottomlessness a floor.

And so Dewey comes forward with *his* solution of the

difficulty. He replaces the idea of separation with the "idea of continuity." Evidently he thinks that it must be one or the other, complete continuity or complete discontinuity. The chessboard must be all white or all black. Resemblance mixed with similarity is impossible. Why? Because they always come mixed, I suppose. We must not admit the most obvious fact of life, because it is contrary to the Gospel according to William James.

"Experience . . . reaches down into nature," cries Dewey; [8] "it has depth. It also has breadth, and to an indefinitely elastic extent. It stretches. That stretch constitutes inference."

Thus experience penetrates nature by being continuous with it. Did you ever try to sew on a button with a needle continuous with the cloth? And think of experience, which is one and indistinguishable from the thing experienced, "stretching" so as to include inference! What inference? If it goes so far as to infer that there is anything to be inferred, it will inevitably stretch this hypothesis until it breaks. The winnowing fan will have blown itself away.

"One can hardly use the term 'experience' in philosophical discourse, but a critic rises to enquire 'Whose experience?' The question is asked in adverse criticism. Its implication is that experience by its very nature is owned by some one; and that the ownership is such in kind that everything about experience is affected by a private and exclusive quality." [9]

There ought to be a law against adverse criticism! But all the Naturalistic Empiricist need do to silence *this* critic, is to point out one bit of experience that has no

[8] *Ibid.*, pp. 4a–1.
[9] *Ibid.*, p. 231.

private and exclusive quality, and is owned by nobody. A toothache, for instance, which aches all by itself, in some public tooth, without hurting a single soul. "The implication"—that an experience has to be owned—"is . . . absurd," Dewey insists.

"It is sometimes contended," on the other hand, "that since experience is a late comer in the history of our solar system and planet, and since these occupy a trivial place in the wide areas of celestial space, experience is at most a slight and insignificant incident in nature. No one with an honest respect for scientific conclusions can deny that experience as an existence is something that occurs only under highly specialized conditions, such as are found in a highly organized creature which in turn requires a specialized environment. There is no evidence that experience occurs everywhere and everywhen. But candid regard for scientific inquiry also compels the recognition that when experience does occur, no matter at what limited portion of time and space, it enters into possession of some portion of nature and in such a manner as to render other of its precincts accessible." [10]

This seems to indicate that the only kind of experience is what we call conscious experience—by which some of us probably mean experience of which somebody is conscious. Certainly it is only experience of this sort which could rightly be called a late comer in the history of our planet. But this gives it precisely that private and exclusive quality that is so "absurd." Evidently, however, Dewey means to suggest—or at least seems to suggest—that experience, since it is "an existence," is something in itself. It "enters into possession," you will note. So I do not own my experiences; they own me. It is not I

[10] *Ibid.*, p. 3a.

who regard them as "slight and insignificant"—indeed I don't!—but it is I who am slight and insignificant in their regard—or maybe large and important. I should like to get their opinion of me, and hope that I shall not be overwhelmed with adverse criticism.

Anyway, experience, when it takes possession of a portion of nature, renders some other portions of nature "accessible." Yes. But accessible to what or to whom? Surely not to any owner of ownerless experience? But this is precisely what Dewey implies. For he goes on to tell about Lyell, who "revolutionized geology by perceiving (*sic!*) that the sort of thing that can be experienced now in the operations of fire, water, pressure, is the sort of thing by which the earth took on its present structural forms." [11] So the things that happened to the earth before experience arrived are made accessible to Lyell by later experiences which he perceived and experienced, but didn't own. Only an adverse critic would ask, "Whose experiences?" and answer, "Lyell's!"

What Dewey wants to say is clear enough. Things happened to the earth before the coming of conscious life upon its surface, and left marks that conscious beings subsequently saw and interpreted. But he is kept from saying this, or indeed from saying anything, by his regard for certain theories utterly inconsistent with the happening or existing of anything.

There are so many absurdities heaped together here that I cannot pause to note them all. In trying to disentangle a skein of yarn which is all tangle, a pair of scissors would perhaps best serve the purpose. But let us get back and learn just how ridiculous Dewey thinks

[11] *Ibid.*, p. 4a.

it is to regard experience as being "affected by a private and exclusive quality."

The "implication is that experience by its very nature is owned by some one." Yes, we remember that. And now he goes on to tell us that "the implication is as absurd as it would be to infer from the fact that houses are usually owned, are mine and yours and his, that possessive reference so permeates the properties of being a house that nothing intelligible can be said about the latter. It is obvious, however, that a house can be owned only when it has existence and properties independent of being owned. The quality of belonging to some one is not an all-absorbing maw into which independent properties and relations disappear to be digested into egohood. It is additive; it marks the assumption of a new relationship, in consequence of which the house, the common, ordinary house, acquires new properties. It is subject to taxes; the owner has the right to exclude others from entering it; he enjoys certain privileges and immunities with respect to it and is also exposed to certain burdens and liabilities. Substitute 'experience' for 'house,' and no other word need be changed. Experience when it happens has the same dependence upon objective natural events, physical and social, as has the occurrence of a house. It has its own objective and definitive traits; these can be described without reference to a self, precisely as a house is of brick, has eight rooms, etc., irrespective of whom it belongs to." [12]

This is the most wonderful house in all my "experience." It is a house that can exist without being owned, it has "properties independent of being owned." Never

[12] *Ibid.*, pp. 231–232.

was there such a house. Excepting of course the Mansions of Philosophy, houses may indeed exist without being owned by men. As a matter of law, houses of that sort are owned by the State—but never mind about that. Let's say that a house may exist without being owned by man. But it must be owned by something—by Nature, and even by Nature's God.

Dewey is a Pragmatist, though he avoids using the word, and yet he here forgets the one sensible tenet ever offered in Pragmatism's name—that there cannot be a difference (say as between a house and a no-house) unless it makes a difference somewhere or somewhen to something or somebody. Did a house exist in no consciousness its existence would make no difference, and therefore would not differ from not existing.

Inevitably, Dewey, having constructed an impossible figure of this sort, finds it exactly fitting to express his philosophy. Because a house doesn't pay county taxes, he assumes that it *is* outside of all mind, all consciousness but its own. Ergo, experience can exist outside of all mind, all consciousness. It doesn't have to be owned. What about Dewey's own egohood? Shall we eliminate that, too? We then shall have the nearest possible approach to the spectacle of a great negation crawling into its hole and dragging the hole in after it which the Unrealist so longs to see. What it means is that he wants to be the great negation himself—the other than anything which can be named. He wants to be God in that state of Pure Being which has not deigned to create. This is what he modestly seeks to describe in his Preface to Genesis.

CHAPTER VIII

THE COLOSSUS

I. PROFESSOR ALEXANDER

PROFESSOR ALEXANDER, Fellow of Balliol and Lincoln Colleges, Oxford, and Professor of Philosophy in the University of Manchester, resembles not a little his famous namesake. If it cannot quite be said that he has no more worlds to conquer, it is at least true that there are none at whose subjection he has not made a valiant attempt. I don't mean that he has scattered himself among details, but that there is something universal, something formidable in his intellectual bulk, something overwhelming and intimidating in the colossal sweep and compass of his indexes and tables of contents.

He was born at Sydney, New South Wales, in 1859, and christened Samuel, but it is difficult to imagine even his intimates calling him Sam. It is significant, I think, that he is always spoken and written of as "Professor Alexander," as though, even in a world where everyone is a professor, he were the professor par excellence.

I confess to a weakness for this transplanted Australian. Here is no narrow provincial with schoolboy prejudices, but—so it seems at first—a philosopher *par métier*. He has a dignity, a severe charm lacking in most of his contemporaries, a readableness which survives the

219

hideous difficulties of his vocabulary and all the endless stretches of his prolixity.

If, in the thought of the day, all roads lead to Einstein, with Professor Alexander they lead further and "postulate an absolute world," a real basis for measurement.[1] He laughs at solipsists because they "could not talk to each other," and seems to snap his fingers—no, that is too lively a figure; say rather he gravely shakes his head in the very face of Immanence—as where he says [2] that the only part of the world "fitted to carry deity" [in an absolute sense] is a part that lies beyond our experience. But be not deceived. Though he frequently spells God with three letters, he never means God without a camouflage of at least five, and tries his best to rob Him of all significance even then.

Not but what, if any mind must set out alone to answer the riddle of the universe, Professor Alexander's is as capable as any. His preoccupations are ethical and aspiring, and those who wish to be chastely pagan might do much worse than to adopt him for their Moses. They need look for no authentic Law, no Sinai, any more than for the bitter sweetness of the Cross. But they need fear no muck-heap. Indeed, there are moments when he seems really God-conscious, when his voice has the roar of thunder and not merely the clatter of theatrically shaken tin—as when he says, "All we are the hunger and thirst, the heart-beats and sweat of God." So "through the thunder comes a human voice." [3] These, however, are but the echoes of an older and (to him)

[1] *Space, Time and Deity,* (New York and London: Macmillan and Co. 1927), Vol. I, pp. 90–91. This monumental work, first published in 1920, contains the substance of the Gifford Lectures delivered at Glasgow in 1916–1918. All references are to the edition of 1927.
[2] *Ibid.,* Vol. II, p. 420.
[3] *Ibid.,* Vol. II, p. 357.

alien thought, Christian sentiments surviving the beliefs from which they sprang—as we shall see.

Nevertheless, he retains—oh, marvel of marvels!—the sweet gift of a genuine humility. In the preface to the 1927 impression of his magnum opus (with which alone I shall deal) he even goes so far as to say: "Whatever doubts I may feel about my work, I do not feel able to offer with confidence any better substitute for it, and I leave it therefore with its imperfections, which I know to be real."

So, instead of doubting the universe, or reason, or the possibility of anybody's knowing anything, he doubts himself and his own work—not too much, but with that half timid, half confident, "There she is, anyway!" of a good craftsman. Everything in the book, he tells us, has been adversely criticized at one time or another, yet he adds that he is "not foolish enough to conclude that therefore the whole is probably right, nor willing to admit that the whole is therefore probably wrong."

An amiable party, this, and doughty. And if you still fancy him incapable of caustic wit, note what he does to Bertrand Russell in the following passage from page vii of the Preface, which I have ventured to simplify by turning into dialogue:

Russell: [4] The importance of the general theory of relativity to philosophy is perhaps greater than its importance to physics.

Alexander: I cannot estimate the justice of that comparison, but of the importance of the theory for philosophy I am sure.

Russell: For my part I do not profess to know what its philosophical consequences will prove to be.

[4] See his *Analysis of Matter,* 1921, p. 55.

Alexander: This is consoling to me, who am in the same case.

Russell: I am convinced that [these consequences] are far-reaching and quite different from what they seem to philosophers who are ignorant of mathematics.

Alexander: Albeit not completely ignorant, I do not [now] feel quite so comfortable.

Great is Alexander the Great! So he can be as sly and deadly as this! Is it possible that Russell's fond habit of going about, hitting people on the head with an empty bladder labeled MATHEMATICS is not always going to be safe?

And now—to get back to our subject—be pleased to read (without any attempt to understand) the following excerpts from Volume I of *Space, Time, and Deity,*—excerpts chosen almost at random, wherein Alexander chants his own faith. Read them aloud, with all the sonority of which you are capable, the book upon a lectern of jewel-studded gold, if possible, in the midst of Gothic architecture and light from stained-glass windows:

"The hypothesis of [this] book is that Space-Time is the stuff of which matter and all things are specifications.

"Space and Time have no existence apart from each other." [5]

"Space-Time is . . . the source of the categories, the non-empirical character of existent things, which these things possess because of certain fundamental features of any piece of Space-Time. These fundamental features cannot be defined. For to define is to explain the nature of something in terms of other and in general simpler things, themselves existents. But there is nothing simpler

[5] P. vi.

han Space-Time, and nothing beside it to which it might
be compared by way of agreement or contrast. . . . The
utmost that we can do is therefore to describe it in terms
of what is itself the creation to Space-Time. . . .

"Space-Time itself and all its features are revealed to
us direct as red or sweet are. We attempt to describe what
is only to be accepted as something given, which we may
feel or apprehend; to describe, as has been said above, the
indescribable." [6]

"Space-Time does not exist but is itself the totality
of all that exists. Existence belongs to that which oc-
cupies a space-time. . . . The world which is Space-
Time never and nowhere came into existence, for the
infinite becoming cannot begin to become." [7]

"Space-Time is in no case a unity of many things; it
is not a one. For that implies that it can descend into the
field of number, and be merely an individual, and be compared as one with two or three. The universe is neither
one in this sense, nor many. Accordingly it can only be
described not as one and still less as a one, but as *the*
one; and only then because the quasi-numerical adjective
serves once more to designate not its number but its infinite singularity. . . . It is not so much an individual
or a singular as the one and only matrix of generation,
to which no rival is possible because rivalry itself is
fashioned within the same matrix." [8]

"In truth, Space-Time is not the substance of substances, but it is the stuff of substances. No word is more
appropriate to it than the ancient one of *hyle*. [Greek for
matter.] Just as a roll of cloth is the stuff of which coats
are made but is not itself a coat, so Space-Time is the

[6] P. 336.
[7] P. 338.
[8] P. 339.

stuff of which all things . . . are made. . . . The stuff of the world is . . . self-contained in that there is nothing not included in it. But it is not the supreme individual or person or spirit, but rather that in which supreme individuality or personality is engendered." [9]

"Space-Time, the universe in its primordial form . . . has no 'quality' save that of being spatio-temporal, or motion. All the wealth of qualities that makes things precious to us belongs to existents which grow within it. . . . It is greater than all existent finities or infinities because it is their parent. But it has not as Space-Time their wealth of qualities, and being elementary is so far less than they are. . . . Space-Time is not in space or time as though there were some enveloping Space or Time. It is itself the whole of spaces and times. . . . But it must not therefore be supposed to be spaceless or timeless . . .

"Call it by what you will, universe or God or the One." [10]

"There is only one Space and one Time. . . . The only eternity which can be construed in terms of experience is infinite time. . . . Space-Time is neither in Time nor in Space; but it *is* Time and it *is* Space." [11]

These are the words of worship. We hear the sound of the cornet, flute, harp, sackbut, psaltery, dulcimer, and all kinds of music. We are moved to fall down before the golden Hyle which Alexander the King has set up. Here is the intricate counterpoint of a later Bach, improvising a vast Doric toccata and fugue. But the more we study Bach the more we are lifted up. And the more we study Alexander—

[9] P. 341.
[10] P. 342.
[11] P. 343.

O Belteshazzar, master of the magicians, because I know that the spirit of the holy gods is in thee, and no secret troubleth thee, tell me the visions of this dream and the interpretation thereof!

2. STEPPING HEAVENWARD

I feel incompetent, even with the help of Daniel, to bring much law and order to these wild, chaotic pastures of Nebuchadnezzar. Moreover, there comes from them a cry, like that of some great truth, mangled and torn. I do not wish to mock it. But could one possibly heal some of its hurts, or at least locate the wounds?

Those who prattle of Space and Time, giving the words that emphasis which suggests capital letters, may usually be understood to speak figuratively. Space, since it gives us the opportunity for action, vaguely suggests Deity, while Time, which we feel as change, stands for the human or receptive end of experience. To emphasize Time is to stress the ephemeral, the individual, the empirical, the subjective—even the emotional. To emphasize Space is to favor the austere, the disciplined, and such principles of constancy as may be felt within or surmised without—but more especially that which is held to be objective and beyond ourselves. But such metaphors usually twist and intertwine in endless confusion.

When Professor Alexander says that "Space and Time have no existence apart from each other," he may mean that we never feel ourselves initiating movements without also feeling time within which to move. When he reduces Space-Time to "motion," he may mean to indicate that motion itself moves upon two legs, one of which (space) remains fixed to mark the progress of the other (time)—which is to say that motion implies a duality. Thus his academic robes have the air of passing unscathed, or torn but little, through the bursting ciphers of Unrealism; of drawing back from Monism, Solipsism, and all nonentity philosophies which would have us believe that the universe springs somehow from one sterile old maid. But he is usually credited with having crowned Time with orange blossoms, giving her in Herr Space a nominal husband only, and leaving her a still unravished bride of loveliness to the end. That is, he is held to be a subjective philosopher, whose Space and Time have no real outward existence. Is he guilty?

Before we can answer we must consider what we mean by subjective and objective. All knowledge, all feeling is of course subjective. It is therefore easy to fall into the error of saying that all that we feel or know is subjective —that the *objects* of feeling are subjective. But a moment's reflection shows the absurdity of such a notion. There is, so to speak, an outside even within ourselves. If I feel happy at the present moment, and know that I feel happy, it is only because I remember that I felt less happy at some previous moment—and this is to appeal to something outside of my present momentary self— using the word self in the sense of "mood."

But apart from the multitude of meanings attached to such words as *I, ego, self,* and the like, the question of the subjective is no more simple than is that of the objective. Personalities are complicated, and sometimes divided—upon the surface. But as we always make use of an inmost self, whatever else we do when it is really we who do it, it would be a good plan to limit the meaning of all such words as far as possible to that which has been subdued and conquered by this inmost self, and brought

beneath its will and sway. What we commonly call the
self is often sadly lacking in unity and harmony. Frequently it is no true follower of the real self, and may
be less obedient to our own than to the wills of others.
Nor is it at all difficult to lessen the depth and breadth
of this inner peace until we become dim in consciousness and much like a weathercock, at the mercy of every
storm of passion and circumstance. But in spite of much
that has of late been written and said (by such writers
as Bergson, more particularly than by Alexander) I
sometimes doubt if going deliberately cuckoo is the way
to health, strength, and felicity.

But the self, however single, is never the all. As Alexander well says:

"A solipsist at one moment could not talk to himself
as he was at a previous moment; he would have no continuous self." [1] Such a solipsist would be unconscious, and
could not exist even save by the grace of something
other than himself—which would rob his solo perform-
ance of its solo quality.

How much more, then, is the objective necessary to
the life of conscious beings who are aware of those con-
cords and discords possible to a duet but incompatible
with a one-voice, a cappella monotone! But it is easy to
see why philosophers shy from space in any sense that
hints at the eternal, for the eternal—whether in lame
tropes we speak of it as being either without or within—
dwells in a "space" that the mind cannot penetrate. It
is unthinkable. Yet we cannot think without it.

Here we are confronted with a concept that, para-
doxically enough, we cannot conceive. I call such concepts
privative concepts, because we can describe only those

[1] *Space, Time and Deity,* Vol. I, p. 90.

qualities of which they are deprived. The Inconceivable is itself such a concept. For obviously we cannot conceive the inconceivable. And yet we can conceive of there being such a thing as the inconceivable. It looms at the end of every vista. We can only describe it by naming any and all conceivables, and coupling them with the word *not*.

A wonderful example may be found in that paraphrase of Romans viii, 29, by Saint Ambrose,[2] where, dealing with the painful subject of predestination, he says: *"Non enim ante prædestinavit quam præscivit, sed quorum merita præscivit, eorum præmia prædestinavit,"*—He did not predestine before He foreknew, but for those whose merits He foresaw, He predestined the reward.

The ticklish part of the proposition is eliminated by that saving *non*. To have affirmed that "He predestined *after* he foreknew," would have been to introduce time into eternity. The *non* removes the whole question from the regions of time and space. It affirms nothing save the denial of all conclusions to be reached by finite comprehension. From how many religious quarrels Saint Ambrose might have saved us!

Professor Alexander was on the right track when he announced that "Space-Time itself and all its features are revealed to us direct as red or sweet are," and bravely added that if we "attempt to describe what is only to be accepted as something given" we attempt "to describe . . . the indescribable." That is, we attempt the impossible when we attempt to describe the *given* in terms of the *not-given*. For language can describe only by suggesting experience. We have, perhaps, tasted the sweet and learned to associate the word with the gusto. We then can "describe" it—but only in terms of sweetness, and

[2] See his *De fide*, V, vi, 83.

only to those who have palates. We know what it is to comprehend much that was at first comprehensible, and so, by analogy, can attach a meaning to the word incomprehensible. We do not comprehend the incomprehensible; but we may, so to speak, comprehend that it is incomprehensible.

But when we are confronted with clean and awful mystery, how different is our feeling from that mental distress which comes from having our noses rubbed in absurdity! Face to face with mystery, men have been known to kneel in silent adoration, and come off never the worse. In fact, to what else may one decently kneel? But upon perceiving absurdity, one must either laugh or go mad. So, if I occasionally permit myself to smile at Professor Alexander, it is only because I wish to remain at large.

Throughout his Space-Time panegyric one catches a sublime refrain addressed to Being as distinguished from existence—as when he says that "Space-Time does not exist." This Being is the supreme quality of Deity, the supreme privative concept of which the human mind is capable, for it seeks to indicate God apart from his works. Pure Mystery. But mystery should not be reduced to mush. "Call it God," says Alexander. And then, just as we are about to comply, we remember that he has described it as "spatio-temporal, or motion." It is therefore but time and space as we know and feel them. It is psychological. It depends upon us. This reduces Supreme Being to breakfast food.

But before turning away from the board, it is only fair to open volume two of the menu and see what is offered for dessert. Alexander has two ways of defining "God." He is, first, "the object of the religious emotion or of worship"; and, second, "the being, if any, which possesses deity or the divine quality." [3]

Now Space-Time might be admitted to be "the stuff of which all things are made," if by that is meant the "stuff" through which they are made *manifest to us*. And it is this interpretation that pops into one's head as one reads, and helps the stuff to go down. But it won't do. Space-Time is here called upon to produce "the being, if any, which possesses deity," to be the matrix in which "the supreme individual or person or spirit"—*if any*—"is engendered." And as Space-Time has been described as "motion," we are asked to believe that it is motion which engenders deity. Try that over your perspicacity! And if you still feel inclined to identify Alexander's "deity" as the Ineffable, or even the Almighty, note the following: "This possessor of deity is not actual, but ideal." [4] We now know what the "if any" means. The philosopher seems to have identified motion (without anything moving) with the Absolute; thought of God the Creator as emerging from the Absolute; and then identifies the Absolute with himself. He means that he, or at most we, make the only God there is.

Is it this not-actual deity, or actual, ideal non-deity, which is the object of his religious emotion or of worship? I really couldn't say. The fact is, I am for the moment incapable of thought, being overwhelmed with an emotion far from religious. Nor does it help to read [5] that "in the religious emotion we have the direct experience of something higher than ourselves which we call God." How can a god made by our ideas, or by those of our animal, vegetable or mineral forebears, be greater

[3] Vol. II, p. 341–342.
[4] Vol. II, p. 353.
[5] Vol. II, p. 352.

than ourselves? At best he could be but the best which
is in us. But it would have helped mightily if he had
said, leaving his theories aside, that in the religious
emotion we have one of the *consequences* of direct experience of something higher than ourselves. This reading
is impossible in view of the text, but it is the only one
that makes sense. For though an emotion is what is felt
of certain subjectively provoked disturbances within the
body, there first has to be an *objectively* provoked (or
remembered) disturbance—as when we see (or remember) a friend, and then experience the emotion of love.
You might say that a religious emotion is what is felt
of this secondary disturbance in a body more subtle than
the physical. But no bodily stir, and no feeling of any
stir, could be higher than ourselves. Only an object could
be that. Not an emotion but a sense is required for experience. It can only be through a religious *sense* that we
could acquire any knowledge of God other than what
we may acquire through the bodily senses. A thrill that we
feel as emotion may, or may not, come as a subsequent
effect. What Alexander here calls God is evidently nothing but a thrill. What caused the motions that caused it,
we are left to guess.

3. IT IS NEVER TO-MORROW TO-DAY

But I cannot bear to drop him and his philosophy at
this sad cross-roads. Too obviously he means to go further. Book IV of volume II, devoted to "Deity," cele-
brates his arrival at his goal. One hears him thundering
forth upon the double open diapason, thus:

"Within the all-embracing stuff of Space-Time, the
universe exhibits an emergence in Time of successive

levels of finite existences, each with its characteristic empirical quality. The highest of these empirical qualities known to us is mind or consciousness. Deity is the next higher empirical quality to the highest we know." [1]

It is a comfort to find a modern philosopher admitting that mind or consciousness has actually emerged from something. But does he mean that Deity, after all, is something actual, now in being, beyond our finite knowledge? Not so. "As actual," he says,[2] "God does not possess the quality of deity, but is the universe as tending to that quality. . . . Only in this sense of straining towards deity can there be an infinite actual God."

"For any level of existence, deity is the next higher empirical quality. It is therefore a variable quality, and as the world grows in time, deity changes with it. On each level a new quality looms ahead, awfully, which plays to it the part of deity. For us who live upon the level of mind deity is, we can but say, deity. To creatures upon the level of life, deity is still the quality in front, but to us who come later this quality has been revealed as mind. For creatures who possessed only the primary qualities— mere empirical configurations of space-time—deity was what afterwards appeared as materiality, and their God was matter. . . . On each level of finite creatures deity is for them some 'unknown' (though not 'unexperienced') quality in front, the real nature of which is enjoyed by the creatures of the next level." [3]

So deity was once matter. In those days, the highest creatures—God knows what created them!—were "mere empirical configurations of space-time." And then, when

[1] Vol. II, p. 345.
[2] Vol. II, p. 361.
[3] Vol. II, p. 348.

appeared "creatures upon the level of life," deity changed and became mind. Would you like to know what is the empirical² quality of a mere configuration of space-time apart from mind? So should I. Alexander, on page 4 of volume I, defines "empirical" as a word "equivalent to experiential." I do wonder what the experience of parallelopipedons was like in those somewhat remote and dim ages when they were the lords of creation—for surely a parallelopipedon is a configuration of space-time, and as empirical as any. And these had matter for their deity.

No, matter did not create them. According to Alexander's philosophy, they must have created matter—assisted, perhaps by isosceles triangles. "Deity is . . . the . . . quality . . . which the universe is engaged in bringing to birth." [4] And at that time the universe consisted of configurations of space-time, and their inferiors—if any.

But while they are engaged in bringing matter to birth, one may pause to admire the neatness with which Professor Alexander has brought spatio-temporal configurations to birth out of his own empirical qualities, then quietly removed himself from the scene, leaving us to suppose that configurations can figure alone without a mind to conceive them—or any deity, either, save a "matter" which they have not yet brought about.

And then, when appeared "creatures upon the level of life," deity changed to mind. And now deity, as something higher still, is *going* to emerge. Heaven speed the day! But such things as have been by my own empirical quality seen to emerge—such things as fish, for instance, growing sportive and leaping out of the water—existed

[4] Vol. II, p. 347.

before they emerged. At least I have always been forced to think of them as existing in some consciousness or other. But Alexander hopes to catch a big fish which shall not exist—nay, shall not be in anywise or anywhere —until he catches it. His deity is simply the next fish. Until it is caught it is nowhere save as a quality in a world that does not have that particular quality. This is certainly being not "actual." And when it is caught it ceases to be deity. "As actual, God does not possess the quality of deity."

Some think that Alexander merely means to suggest that God impinges upon our experience only when he becomes manifest in one manner or another. But such an explanation is totally at variance with his main contention that deity has no being whatever save in that level below deity which is to experience the manifestation. The word manifestation (which he does not use) would be absurd in such a connection. There is nothing to be manifested, nothing to be experienced, according to the hypothesis. The lower level rises into a higher level which the lower creates in rising to it. The lower encounters itself just ahead of itself!

But whose are the empirical qualities of which we now are to hear? I fear they again are Alexander's. For a deity that is but an ideal would be hard put to it to have experiences. And as for a collective mind, Alexander says, rightly enough,[5] "There is no sufficient evidence that such a mind exists." But let us not press the point. Here is a still better one:

"If Time were, as some have thought, a mere form of sense or understanding under which the mind envisages things, this conception [of deity continuing to crop out

[5] Vol. II, p. 241.

from Space-Time] would be meaningless and impossible." [6] At the risk of having an unreal Space-Time deity become impossible, you must find a Time flowing apart from any sense or understanding, God's or man's. I wish you joy of your search. If and when you succeed, you will discover that Time "has no special relation to mind," and that "mind . . . is but the last complexity of Time." The last as yet.

It may, however, console you to learn [7] that "should the extension of mind beyond the limits of the bodily life be verified, so that a mind can either act without a body or may shift its place to some other body and yet retain its memory, the larger part of the present speculation will have to be seriously modified or abandoned." But—unless we *do* indulge in a little modifying or abandoning—

"Bare Time in our hypothesis . . . will be completed by the conception of God." [8] Clearly he means by *our* conception of God. For "Bare Time is the soul of its Space, or performs towards it the office of soul to its equivalent body or brain." [9]

Don't ask *me* to explain this! Maybe Bare Time was the soul of Empty Space. Anyway, it seems to be suggested that Time secretes Space even as the soul secretes body and brain. Just about, I should think. By giving Time this "office" we clearly elect it to the presidency. "And this elementary mind which is Time,"—the same Time, I suppose, which "has no special relation to mind,"—"becomes in the course of time so complicated [with what?] and refined in its eternal groupings that there arise finite beings whose soul is materiality, or color, or life, or in the

[6] Vol. II, p. 345.
[7] From Vol. II, p. 424.
[8] Vol. II, p. 345.
[9] Vol. II, p. 346.

end what is familiar as mind . . . Time is the principle
of growth and Time is infinite." [10] So Time secretes not
only space, but souls—even colorful, material ones—as
well as minds. Really, it would be too bad to have to
modify or abandon such a speculation as this.

But if, to continue the speculation, "Deity is . . . the
next higher empirical quality of mind," and if "the universe" be really "engaged in bringing" it "to birth," naturally "the universe is pregnant with such a quality."

But "what quality is we cannot know . . . If we could
know what deity is [he must mean if we could know what
the next deity is going to be!], how it feels to be divine,
we should first have to have become as gods . . . its
nature we cannot penetrate. . . . It is fitly described . . .
as the color of the universe. For color . . . is a new
quality which emerges in material things in attendance on
motions of a certain sort."

Imagine the *chromatic* without a sense of sight also "in
attendance"!

"I am leaning for help," he says, "upon [George]
Meredith, in whose *Hymn to Color,* color takes for the
moment the place of what elsewhere he calls Earth: a
soul of things that is their last perfection; whose relation
to our soul is bridegroom to bride." [11]

But it will be remembered that Meredith wrote of color:
"Thy fleetingness is bigger in the ghost than Time with
all his host." So Professor Alexander makes haste to add:

"As I read the poem, deity as the next higher empirical
quality is not different from color as he conceives it; save
only that for him the spirit of the world is timeless,
whereas for us deity is, like all other empirical qualities,

[10] Vol. II, p. 346.
[11] Vol. II, p. 347.

a birth of Time and exists in Time, and timelessness is for us a nonentity."

Which is to say that Alexander leans on Meredith not by leaning on him, but by removing the very prop upon which Meredith leans. But we are getting the color of Alexander at last. He worships something—a God which he feels but does not know. And could he but come to grips with his thought, and say that it is our *knowledge* of God, that is empirical and conditioned by all our limitations, he would at once break from the ranks of the Unrealists. But he cannot. He makes God the creature of our experience, of *our* empiricism, and thus is caught in the meshes of the Greater Silliness—or is it the Lesser? Lilliputian threads seem to hold our Gulliver down. It is only his desire, his yearning, which is great. He is like Reuben Light, whom Eugene O'Neill so lately made to worship a dynamo—and first made mad, it seems to me, chiefly to save the author from the charge of taking seriously any of his creature's incredible ravings.

But Professor Alexander is not mad—unless it be north and by west. The most we can say of him is that his mind has become too complicated and refined in its eternal gropings. He worships Time, but not the present time, nor the past. God is the Future. The universe is engaged in bringing it forth. So the universe (God help us!) is the mother of God. The universe is pregnant with a new color, which will emerge from or in material things in attendance on motions of a certain sort. The universe seems to have been impregnated by motions of a certain sort. Meredith "figures the relation of our soul to color under the metaphor of love." And "color takes for a moment the place of what elsewhere he calls Earth." [12] So color, elsewhere

[12] Vol. II, p. 347.

called Earth, and now metaphorically expressed as love, is the fetus which swells the belly of Space-Time. Will it be God when it is born? Is God, after all, going to be Love? Not *this* god. When he is born he will become a human concept, like color—love in our own hearts—purely subjective. He is in Space-Time, in the present or in the past of "motion," or else nowhere at all. And motion is held to have once existed without the help either of mind or matter, a motion that must have been the same as rest.

And yet, every now and then there comes from these pages a flash, not so much of color as of light—very feeble, very inconsistent with the general darkness, but real. This Space-Time deity, before birth at least, is a mystery. This motion which is at rest may be itself a metaphor, born of the sense of Pure Being. If one cuts one's way through the Alexandrian maze in this fashion, one arrives at a sensible thought—the Absolute conceived as something different from all that is positively conceivable. But it is possible to read sense into *any* philosophy in this fashion. For the most part Alexander's deity is simply To-morrow—a to-morrow which, to-morrow, will still be —To-morrow. It is, as he says, "an ideal." This limits it to human comprehension. His incomprehensibility is but the incomprehensibility of contradictory language.

At the same time, Alexander does entertain a God, a most forceful God, more especially when he denies him most, calling him by many names, always trying to keep Him out of sight, as if He were a poor relation. Did you notice how, a few pages back, he said that there could be an "infinite actual God" only in the sense of a universe "straining" towards deity? How he then defined God as "the principle of growth"? How he asked us to envisage

space and time as something beyond sense or understanding—which we could not do because of his injunction to
envisage them at the same moment as spatio-temporal?
Real Deity lurks behind these words like "straining" and
"growth." The universe is pregnant with it—and force
hides within this pregnancy like a lion in an ass's skin.
Only, Alexander insists upon translating the roar as a
bray. He cannot bear the thought of Cause, and confuses
it with motion, which is but a result of cause.

So he says [13] that "all causality" is "the continuous passage of one motion or set of motions into a different one."
He avoids the word "causation," and this makes it possible to suggest that "causality," which is the relation of
cause and effect, may be evoked without implying the
cause end of the relation! So he defines causality as the
passage of one motion into another. That is, it is all effect.
This is a Bertrand Russell subterfuge. True cause, on the
contrary, is merely incomprehensible. But instead of
attaching the idea of mystery to this, which is the real
and living mystery, he limits that awful quality "which
plays the part of deity" to the effects that Deity is to have
upon ourselves. Cause is ignored.

For Professor Alexander is one of those philosophers
who think that to confess that God is already God, is to
impose an intolerable limitation upon ourselves. It makes,
they say, the universe already at an end when it begins.
It leaves us, you see, only an infinite field for progress,
and when we have infinitely progressed we shall be
through. I find this notion implicit in almost all of the
Unrealists. And no wonder. For they first make infinity
finite, humanly narrow, out of themselves. They are so

[13] Vol. I, p. 283.

"empirical" that they fail to note that it is their own experiences which are in the making, not Deity who is on the make.

But, "since Time-Space is already a whole and one, why, it may be urged, should we seek to go beyond it?" That is, even into the future. "Why not identify God with Space-Time?" [14] Why not take her as she is, alter the ancient litany, and pray: *We beseech thee to hear us, good Space-Time!* And the reason given is that "no one could worship Space-Time." I am inclined to think that here Alexander is right. This Space-Time is only the past up to the tick, and it certainly would be difficult to cry with any fervor: *Present* Time, have mercy upon us! Past Time, have mercy upon us! As he says:

"It [Space-Time] may incite speculative or intellectual admiration, but it lights no spark of religious emotion. Worship is not the response which Space-Time evokes in us."

"In one way," he goes on, "this consideration is irrelevant; for if philosophy were forced to this conclusion that God is nothing but Space-Time, we should needs be content. But a philosophy which left one portion of human experience suspended without attachment to the world of truth is gravely open to suspicion; and its failure to make the religious emotion speculatively intelligible betrays a speculative weakness."

True enough. It is not altogether irrelevant to speculation that we are unable to worship our own knowledge. So Alexander turns to the time to come—though he bids it be created out of the past, thus putting all power into the hands of the gone. I am willing to admit that power acts

[14] Vol. II, p. 353.

through the past, but doesn't that make it a present power? Not for Alexander, who always puts power in the direction in which he is not looking. He cannot get out of the barber-shop. His deity is the great Next!

"Of such a thing," he says, "the whole world is the 'body' and deity is the 'mind.' But this possessor of deity is not actual but ideal." This jumble of words, however, does not satisfy him. He must add: "As an actual existent, God is the infinite world with its nisus towards deity." [15]

The only solvent credit in this declaration of bankruptcy is now called Nisus, a word meaning "effort," blood brother to "strain." Alexander constitutes himself its hierophant, and speaks of it again and again—as on page 348 of this same Volume II, where he talks of "that restless movement of Time, which is not the mere turning of a squirrel in its cage, but the nisus towards a higher birth." And on page 367: "Deity is a nisus and not an accomplishment." And on page 367: "The nisus in the world which drives it, because of Time, to the generation of fresh empirical qualities is a verifiable fact."

Here "time" grasps the throttle with that thoughtless "because," but in general Nisus hath all power and glory. Nisus performs all the physically creative functions of a God. It is, then, but another name for God, in this narrow aspect.

I can understand why a man should wish to find another name for God. The word has been so long abused by those wishing to give divine authority to their own opinions that a certain odium attaches to it. By all means say Nisus, if you know and admit what you mean. But that is just the trouble. Nisus denies what you mean. It is like those words

[15] Vol. II, p. 353.

"tendency" and "evolution," which are always being invoked to perform miracles, and then—the miracles having been performed by something—are held to prove that it forcelessly came about. Such words have a concealed significance. On the other hand, great words, like God and Deity, are sometimes robbed of their significance—and in such cases I admit to a tendency to treat them flippantly, as becomes their flippant use, having a certain horror of being thought to revere the inanity of the author's intent. As to Nisus, it very often raises the suspicion that it is but a more chaste way of spelling Bergson's *élan vital,* which also serves to introduce the Stranger who is in the house without granting him his nature and quality. If the truth which Nisus is made both to employ and to conceal were openly proclaimed, what a different philosophy Alexander's would be!

And yet, it is in its way a noble philosophy. It makes no authentic appeal to the head, but it does appeal to the heart. Moved by the insight of love, the Colossus feels some reality beyond the self. But he must include it in Space-Time, and can find no sound philosophical basis for the externality which he occasionally attributes to it.

This "God" which he offers to us as an object of "worship,"—to us, who, "if we are to follow the clue of experience," he says, must believe that "the claim for the future life is founded on error—" this god is only the god of next week, or of Friday fortnight, who, when Friday fortnight shall have rolled round, will be but common clay.

"Thus there is no actual infinite being with the quality of deity; but there is an actual infinite, the whole universe, with a nisus to deity; and this is the God of the religious consciousness." [16]

[16] Vol. II, pp. 361–362.

It is really disquieting to think what is being taught in our Universities. And I have but one more "next" to offer —Professor Alexander's noted disciple, Alfred North Whitehead.

CHAPTER IX

COLOR OFF ON THE WALL FOR US

I. WHITEHEAD

"WHATEVER you do in your next book, be sure and give Whitehead hell," writes a correspondent.

It is a tempting suggestion. For with Whitehead as with Russell, there are no things or substances in the ordinary sense, but only events—an event being "the grasping into unity of a pattern of aspects." Nor does he especially endear himself to me when he says that "the effectiveness of an event beyond itself arises from the aspects of itself which go to form the prehended unities of other events." [1] It may be so, but one wonders if something could not be done about it.

An event is an "epochal occasion," and—but why go grasping through all this "Unity" again? It begins by "presupposing the organic theory of nature," and ends by claiming to "have outlined a basis for a thoroughgoing objectivism." [2] The weary mind—which perhaps does not crave an objectivism quite so thoroughgoing—lies down in the traces, its curiosity destroyed by such thoroughgoingly damnable language. It surmises already that "God" will be "immanent," that each "creature" will "concretize" the past and most of the future in a universe

[1] *Science and the Modern World* (New York: The Macmillan Co., 1928), p. 174.
[2] *Ibid.*, pp. 173–174.

whose objectivity is subjective and confined to the first person singular. Frankly, I shrink from the task of interpretation, and feel inclined to let the hells of Russell and Alexander suffice for the sins of Whitehead.

But the fact is, they more than suffice. If at times it is difficult to distinguish him from Alexander more especially, yet it soon becomes apparent that he has his own peculiar salvation. So let us turn to this—surely the more promising subject.

To began with, Alfred North Whitehead is the greatest mathematician of all the modern philosophers. This is not to compare him with Einstein, or any other of the great professionals who are amateur only when it comes to philosophy, but with those who are amateur when it comes to mathematics. This is a point worth mentioning, even though his salvation, such as it is, comes from quite another quarter.

He was born on the Isle of Thanet, the son of a canon, February 15, 1861, and was educated at Trinity College, Cambridge. As he was Professor of Applied Mathematics at the Imperial College of Science from 1915 to 1924 (when he accepted the chair of philosophy at Harvard), was President of The Mathematical Association in 1915-1916, and has held the post of Senior Mathematical Lecturer at Trinity, it may seem strange to speak of him as an amateur. But I only mean that he is, as to mathematics, a scholar rather than an inventor, and interested chiefly in the philosophical implications that he is able to draw either from mathematics or from any other study.

It would be absurd to suggest that his gifts are not of a high order in nearly all directions. To treat him at all adequately would require a book—an almost incomprehensible book. So I propose to offer only a few pages, dealing with but one and that the smallest of his works, *"Symbolism, its Meaning and Effect,"* which first took form as the Barbour-Page Lectures, delivered at the University of Virginia in 1927.[3]

Imagine yourself in the awful situation of a student at one of these lectures, and listening to the following:

"Our experience, so far as it is primarily concerned with our direct recognition of a solid world of other things which are actual in the same sense that we are actual, has three main independent modes each contributing its share of components to our individual experience. Two of these modes of experience I will call perceptive, and the third I will call the mode of conceptual analysis. In respect to pure perception, I call one of the two types concerned the mode of 'presentational immediacy,' and the other the mode of 'causal efficacy.' "[4]

"We [should] distinguish that type of mental functioning which by its nature yields immediate acquaintance with fact, from that type of functioning which is only trustworthy by reason of its satisfaction of certain criteria provided by the first type of functioning. I shall maintain that the first type of functioning is properly to be called 'Direct Recognition,' and the second type "Symbolic Reference.' "[5]

"I shall argue on the assumption that sense-perception [presentational immediacy] is mainly a characteristic of more advanced organisms; whereas all organisms have experiences of causal efficacy whereby their functioning is conditioned by their environment."[6]

[3] Published by the Cambridge University Press and by The Macmillan Co., New York, 1918.
[4] P. 19.
[5] P. 8.
[6] P. 5.

I have given these quotations in inverted order, but don't let me keep you from reading them the other way around. Evidently Whitehead believed with Mark Twain that "no real gentleman will tell the naked truth in the presence of ladies." One mustn't if one wishes to earn a large salary and to sit in a chair of philosophy. But I, who have no ambitions, will venture to tell what he means but doesn't intend to be caught meaning, not even by himself.

Causal Efficacy refers to the Outside, the Out There. Presentational Immediacy refers to the Inside, the In Here. One implies the other; from one we infer the other. This inference is called Symbolic Reference, because in the whole more or less English language there could be found no expression more obscure to apply to the matter in hand. Presentational Immediacy is held to belong only to the higher organisms because it amounts to conscious perception. But all things, even sticks and stones, are played upon by Causal Efficacy—an efficient cause outside of themselves. Causal Efficacy, therefore, reaches ultimately to God.

As an expresion for describing that part of experience which depends upon the capacity of the receiving end, Presentational Immediacy is the worst ever, for it suggests that the self, the will, the soul, is something less than a concrete existent, and seems to reduce it not so much to that which is presented to it as to the immediateness of that which is presented to it—which is like saying that a man who keeps his engagements consists of his promptness.

The words Causal Efficacy are almost equally open to objection, since they deny to Cause the dignity of a noun. "Efficacy" the only noun employed, is but the abstract quality to be derived from the fact of an effect being effective. The language is made as thin as possible, and as sheer thinness I think it deserves the Nobel Prize. Nevertheless, it represents the sincere struggle of an able mind to get out of the hole which the influence of William James more than any other digged for philosophy with his denial of that "inner duplicity" of which so much has been said.

Both Whitehead and Alexander are undoubtedly voices crying in the wilderness, the forerunners of a new trend in philosophy. But, like all representatives of transition periods, they have accents that are faltering and confused. Whitehead, for example, seems afraid to say anything without deleting it as far as possible of all meaning. Even what he does say is drowned in the shrieks of the tortured syllables with which he says it. Possibly the fanfare is designed for that very purpose, and certainly his literary style contributes to that contempt combined with sense-less awe which the average citizen feels for all that passes for "highbrow."

But I respect him so much that I have dared to go beyond the letter of his law, and to draw out a meaning which he makes felt rather than intends. He would not agree with the interpretation which I have just put upon his paragraphs. But leaving that for the moment, let us, while continuing within the lecture-hall, cast an eye upon the chair—the literal chair—that the speaker upon the platform vacated when he rose to address us.

2. THE CHAIR AND THE PUP

"We look up and see a colored shape in front of us," he says,[1] "and we say,—'There is a chair.' But what we

[1] P. 3.

have seen is the mere colored shape. Perhaps an artist might not have jumped to the notion of a chair. He might have stopped at the mere contemplation of a beautiful color and a beautiful shape. But those of us who are not artists are very prone, especially if we are tired, to pass straight from the perception of the colored shape to the employment of the chair, in some way of use, or of emotion, or of thought. We can easily explain this passage by reference to a train of difficult logical inference, whereby, having regard to our previous experiences of various shapes and various colors, we draw the probable conclusion that we are in the presence of a chair. I am very sceptical as to the high-grade character of the mentality required to get from the colored shape to the chair.

"One reason for this scepticism is that my friend the artist, who keeps himself to the contemplation of color, shape and position, was a very highly trained man, and had acquired this facility of ignoring the chair at the cost of great labor. We do not require elaborate training merely in order to refrain from embarking upon intricate trains of inference. Such abstinence is only too easy.

"Another reason for scepticism is that if we had been accompanied by a puppy dog, in addition to the artist, the dog would have acted immediately on the hypothesis of a chair, and would have jumped on to it by way of using it as such. Again, if the dog had refrained from such action it would have been because it was a well-trained dog.

"The transition from a colored shape to the notion of an object which can be used for all sorts of purposes which have nothing to do with color, seems to be a very natural one; and we—men and puppy dogs—require careful training if we are to refrain from acting upon it." [2]

Professor Whitehead has stumbled over this chair, I think, and thrown himself headlong. Perhaps he was *very* "tired," otherwise he would hardly have permitted himself to "pass" from the "perception of the colored shape" to the "employment of the chair" in any such way of "use, or of emotion, or of thought." He is "sceptical as to the high-grade character of the mentality required to get from the colored shape to the chair." But there is no getting from the shape to the chair. Not even the puppy dog gets as far as that. He merely gets from a visual sensation to the anticipation of a touch sensation, which he then proceeds to acquire. Experience has taught him that if he sees a chair, or has that experience which we call seeing a chair, he has but to indulge in certain motions—say a leap—and he will straightway have the yet more charming experience which we call feeling a chair (peradventure a nicely cushioned one) beneath his buttocks.

Whitehead speaks here (and in many other places) as if there was something more real, more chairish, in feeling a chair than in seeing a chair, as if sight were a mere nothing and touch brought us the real goods. True, touch is the more fundamental sense, and older in the history of the world. But matter is not any more real in being hard than it is in being colorful. The puppy dog passed from sight to touch, not from color to chair. But first, and far more important, he passed from present experience of one sensation to hope of another. The fact that, after he had put his hope to the test, experience shifted from one sense to another, is of no consequence. He might have stumbled against the chair in the dark, and

[2] P. 4.

founded his hope upon a bruised nose, thus sticking to touch all the while. The noteworthy leap was from the present to the future.

And Whitehead is much mistaken if he thinks that a leap implies no training. It is far from "natural." A baby will creep straight out of an open French window if left to itself. It is not born knowing what will probably support its weight and what will not. Its first hypothesis is that the green-colored shape which presents itself to its eyes (and that its elders call the lawn) twenty feet below, is an extension of the carpet. Only a well-trained baby—trained by the good luck of having received a number of sound bumps before encountering a fatal one—will entertain the hypothesis called space or air.

The training to which Whitehead subjects his dog and his artist is really retraining, or untraining. The artist overcomes his acquired habit of thinking of certain touch sensations that have long been associated with sight, and sticks to sight. He returns as far as possible to the more primitive, and lives in the present as regards the chair —though in regard to his picture I trust he is somewhat more sophisticated. And I do hope that he has not read so much Whitehead that he begins to doubt that what looks like a chair would probably feel like a chair if he made the necessary adjustments. Is the unfortunate pup (I refer to the dog) so trained that he no longer believes that cushions would feel soft if put to the way of use? My dog sometimes refrains from cushions, but I have always been under the impression that he does so because he entertains the hypothesis of a possible slap on the snout. For, the moment my back is turned . . .

However, there remains Whitehead's scepticism "as to the high-grade character of the mentality required to get from the colored shape to the chair," which we have just seen is a getting from a present experience to a belief in a future one, an indulgence in inference, in short.

He seems to prefer a mentality incapable of inference. Why? Because, though it is very difficult to refrain from inference once you have acquired the habit, it is easy not to acquire the habit. He admires artists stout enough to unlearn what they have learned—and so do I, if that learning stands in the way of their peculiar vocation. And still more he ought to admire one who couldn't be taught anything in the first place—one who would continue to walk right into patches of color to the end of his days, and always be surprised whenever he barked his shins. But this is like admiring one who doesn't know enough to come in when it rains.

Whitehead certainly didn't intend to go as far as to say this. No less certainly, he has a grouch against inference, more especially that naïve sort which he calls Symbolic Reference. It is the sort employed by the practical man, who, having discovered that seeing a chair generally means a chance to sit down, lets it go at that. If he turns philosopher and begins to ask what it all means, and if a chair remains a chair in the dark when there is nobody sitting in it, his inferences become what Whitehead calls Conceptual Analysis. But neither sort gets much of a hand from Whitehead. They smack too much of thought; and though he is a metaphysician himself, he is somewhat infected with the modern distrust of Mind.

Now I should say that if this dog who entertained the hypothesis of a chair entertained the hypothesis that his chair experiences came from something which was not dog; if he believed that he felt, saw and smelt chairs because, crudely speaking, there *were* chairs; he was a most philosophical dog, and (far from being dismissed as untrained) should have been made an F. R. S., an Sc. D. (Cambridge), and otherwise mentioned in despatches. He had achieved the feat of making a symbolic reference from his own presentational immediacy to that older causal efficacy without. Is he to be set aside in favor of that retrained brute who no longer knew the emotional use of chairs when he saw them?

But Whitehead would have it that both causal efficacy and presentational immediacy are "modes" of "direct recognition," requiring no thought, no reflection. He is properly impressed by the fact that experience could not take place without something to be experienced as well as something to experience it. But he is so afraid of losing hold of this outside end of the business that he tries to make even our "recognition" of its two-endedness a matter of direct perception. This is why he would not agree with my suggestion that causal efficacy is the same thing as efficient cause.

"This symbolism from our senses to the bodies symbolized," you see, "is often mistaken. . . . A cunning adjustment of lights and mirrors may completely deceive us." [3]

Alas, yes! It is perfectly certain that when we have a sensation we have whatever sensation it is that we do have. But it is by no means certain that when we go trustingly forward in search of anticipated sensations we shall not fall into painful, or even delightful, surprises. I knew a man once who married a helpless, weak, and clinging woman—and he was simply amazed. But nothing risk, nothing have. Our knowledge of Cause can never be perfect because our experience of Effect, of what Cause

[3] P. 5.

may do to us, is not infinite, nor our memory and judgment above reproach. At the same time, there is no reason for doubting that there was a Cause that did whatever it did. Doubt comes in only when we began to speculate upon what is in store.

Whitehead's confusion of thought is exasperating and unpardonable. There is no end to his argumentative legerdemain. For example, if we turn back to page 2, we find him discussing that "sort of language . . . which is constituted by the mathematical symbols of the science of algebra."

"In some ways," he says, "these symbols are different to those of ordinary language, because the manipulation of the algebraical symbols does your reasoning for you, provided that you keep to the algebraic rules." But "you can never forget the meaning of language, and trust to mere syntax to help you out."

Unquestionably the rules of mathematics may be made to do some of your thinking, and so may an adding-machine, but only in the sense of enabling you to make use of somebody else's thinking without taking the trouble to do it over again. There is, to cite an example, the rule which says, "If you transpose a symbol from one side of an equation to the other, you must change its sign, so that plus becomes minus and minus plus." Remembering this, I can say that if x plus y equals z, then x equals z minus y, without stopping to realize that I have in effect taken the first equation and then maintained its equality by subtracting y from both members of it. But the man who made this rule did some thinking, I fancy; nor should I get very far in mathematics if I forgot the meaning of the signs, and couldn't distinguish plus from minus.

Perhaps, in speaking of "recognition" as "direct," as

something superior to reflection (as if we could re-cognize without reflecting), Whitehead was, just this once, permitting syntax to do his thinking for him. He forgot the meaning of language. Everybody is doing it. So let's help him out by giving him another chance.

This time we shall, if you please, station ourselves in front of a wall, say one of the walls of the lecture-hall, though I myself should prefer an outside view.

"The experienced fact is 'color on the wall for us.' Thus the color and the spatial perspective are abstract elements, characterizing the concrete way in which the wall enters into our experience. They are therefore relational elements between the 'percipient at that moment,' and that other equally actual entity, or set of entities, which we call the 'wall at that moment.' But the mere color and the mere spatial perspective are very abstract entities, because they are only arrived at by discarding the concrete relationship between the wall-at-that-moment and the percipient-at-that-moment. This concrete relationship is a physical fact that may be very unessential to the wall and very essential to the percipient." [4]

If you feel a little "tired" now, don't remain standing, but pass at once, like an untrained dog, from chair-as-seen to chair-as-felt. The ice-water is right there on the table in all its casual efficacy. You may be very unessential to it, but it may be very essential to you as percipient.

The experienced fact (presentational immediacy) is color on the wall for us, and he elsewhere indicates the externality (causal efficacy) of the wall itself as "color off on the wall for us." So that is settled. But what he is trying to prove is that we have "a direct experience of an external world,"—truly an admirable ambition. And he

[4] P. 18.

refers us to "the first portion of Santayana's recent book, *Scepticism and Animal Faith,* for a conclusive proof of the futile 'solipsism of the present moment,'—or, in other words, utter scepticism—which results from a denial of this assumption." [5]

But we need no proof of the futile and utter scepticism that results from denying that we have experiences of the external world. What we need is the proof of reason, not animal faith, to convince us that the very fact that we have experiences of any sort is sufficient evidence of an external world.

Whitehead, however, here seeks to prove it by saying that the relationship between the percipient and the perceived is a "physical fact." An actual fact, yes. A physical fact, no. If ever there were a mental fact it is this. How can a relationship be physical? We conceive of the idea of relationship by a pure act of reasoning. But Whitehead would have it that relationship is not a concept but a percept. Can we, then, see a relationship? Or do we touch it, or hear it, or taste it, or smell it? After this, we need not wonder at this saying.[6] "It is a matter of pure convention as to which of our experimental activities we term mental and which physical." It must then be a matter of pure convention as to which of my experimental activities I term my own, and which I term not my own. Whitehead's distinction between presentational immediacy and causal efficacy suddenly disappears.

And yet he is in the main beautifully alive to the real potency of the subject and of the object, as when he says [7] that "every actual thing is something by reason of its

[5] Pp. 33–34.
[6] P. 24.
[7] P. 31.

ctivity." And [8] "we must conceive perception in the light of a primary phase in the self-production of an occasion of actual existence. In defence of this notion of self-production arising out of some primary given phase," he continues, "I would remind you that, apart from it, there can be no moral responsibility. The potter and not the pot is responsible for the shape of the pot."

This is not Unrealism, it is sense. "Self-production of an occasion" is merely an unhappy way of referring to the part played by the will in all conscious experience. "In this way we assign to the percipient an activity in the production of its own experience." [9]

Certainly we do. If we were not something we could not have experiences. But what if I were to say, on top of this, that it is a matter of pure convention as to where the pot begins and the potter leaves off; that there is no way of distinguishing the potter from his clay?

But I go back to the wall, and ask what he means by saying that our relationship with it may be important to us but not important to the wall? Perhaps he wishes to suggest that it is more important to us that we perceive our environment than it is to our environment to be perceived. Often this must be the case. But what about the following:

"The spatial relationship is equally essential both to the wall and percipient; but the color side of the relationship is at that moment indifferent to the wall"? [10]

And this; [11] "An immediate presentation of an external world is in its own right spatial"?

Here his old faith in touch and his distrust of sight

[8] P. 10.
[9] P. 10.
[10] P. 18.
[11] P. 28.

once more invade his philosophy. He means that space is perceived, and that therefore there is no doubt about it. But I fear that space is just as indifferent to our perception of it as is color. Nor is it at all possible to think of either color or space as having rights apart from all perception. Color and space as we know them are color and space as we feel them. If nothing feels them they lose all their spacious and colorful character.

Thus when he says that "direct recognition is conscious recognition of a percept in a pure mode, devoid of symbolic reference," [12] I can only perceive pure syntax. Of course our perception is the result of the fruitful embrace of self and not-self. But by the time I have consciously recognized it as such I have destroyed the precious purity of the mode—and willingly. I refuse to abstain from my symbolic reference, my thought. Whitehead aims at something which "does away with any mysterious element in our experience." [13] Oh, what a tangled web we weave for ourselves when once we practice to deceive ourselves. From this undoubtedly comes his mad endeavor to put the mystery into words. And yet, though they emerge from the attempt quite meaningless, his fundamental impulse is sound. He would win for us again the right to believe in reality. He doesn't quite succeed, but perhaps we ought to be thankful even for that little bit of color off on the wall for us.

[12] P. 22.
[13] P. 11.

CHAPTER X

THE RELATIVITY OF MORALS

IF Luther, when he threw his ink-pot at the devil, had only *hit* him! But, judging by the system of philosophy just reviewed, his aim, if not his intent, appears to have been bad. So James, Einstein, and their followers have among them succeeded in abolishing (theoretically) all distinctions between good and evil—James by his denial of real distinction anywhere, and Einstein by holding every basis of measurement or judgment to be an arbitrarily adopted *fils d'amour*.

Neither of these gentlemen wished to do harm, or had very radical ideas as to what sort of conduct is moral. But if their fundamental doctrines are to be accepted, the world is going to be at its wit's end trying to discover rational grounds for a moral code; while as for the chances of enforcing it, even a little, they resemble the notorious "Chinaman's chance" in San Francisco in the days of Dennis Kearney.

How would you like to find yourself teaching a class of children that honesty is the best policy, and be compelled to say:

"Of course there is no such thing as stealing, really, because there is really no such thing as a place. Therefore it would be impossible to take anything out of a place called somebody else's pocket, and put it into another called your own. Nor is there any duplicity in the primary

stuff, such as is implied by the words *meum* and *tuum*
But we will pretend that there is, and pretend that there
are places and pockets and straight lines, even the line o.
rectitude.

"The great objection to stealing, or to anything else
which is forbidden, is getting caught at it. That is, such
is the objection from your own personal points of view
Society at large punishes certain conduct because it is
considered detrimental—to and by society at large. But
your share in society at large is very small. You will
scarcely notice the trifling addition to the public burden
made by anything which any one of *you* can do. And if
you are not troubled by a conscience—and I don't see why
you should be—you will each and every one of you be
clear gainers by anything which you can—eh—appropriate without detection. But you must use good judgment, and not overrate your individual talents in this
regard.

"What I am trying to do is to condition your reflexes
so that you will have proper reactions and blindly refuse
to do anything which is called wrong, even though you
know it is purely arbitrary to call anything either wrong
or right. This is called the social ideal. Of course I cannot point out anything which is noble, or for that matter,
ignoble, in this or any other ideal, except in relation to
some standard of conduct which is chosen for the public's
supposed convenience and constitutes the very ideal in question. However, I expect you to act just as nice little boys
and girls used to act in the dark, superstitious ages when
they believed in real nobility and in real distinctions. In
fact, as the world has progressed so much since then, I
expect you to act considerably better—though why you
should, beats me.

"The next class I teach I am going to pretend that I
really believe in right and wrong, and take my pupils when
they are so young that by the time they learn better it will
be too late—though why I should I can't see, either. I much
doubt the actual existence of other people, and it can
hardly make any difference to me how they act. Even considering myself as a member of society, my personal gain
or loss will be very little from anything that I can accomplish for society. The sacrifice amounts to more than the
possible dividend. Besides, children who are not fools will
soon see through me and follow my example rather than
my precept, and so become unconditioned, or reconditioned.
Let other teachers try it if they like. *This* school is dismissed."

Do the pundits of Unrealism hold forth in this fashion?
They do not—at least not many of them. On the contrary,
when they come to discuss one's duty to one's neighbor,
they give the impression of having a but-slightly-edited
Sermon on the Mount beside them. The fact is, they do
have it beside them, or not far off. Its echoes are in the
air they breathe. Or at least the echoes of some code
founded upon a belief in the supernatural. What a modern-
istic morality would actually be like in practice we have as
yet no means of knowing—except perhaps in Russia. It
is all theory thus far. And though signs are not wanting
which indicate that its chance is coming, it can hardly be
said to have been tried. The teaching which is now going
on in our colleges is merely sowing the wind.

For it cannot long remain a secret that the high-sounding
words with which Unrealists usually close their volumes,
are totally inconsistent with the philosophy of the three or
four hundred pages preceding them. These moral codes
are still colored by ancient sentiments, and by practical

considerations. An Unrealist *should* have *no* theory of
ethics. Nevertheless, these tail-pieces and interludes are interesting; and for our first sample we may as well go back
a little, say to Herbert Spencer, the great Cham of Hedonism, or the doctrine that good conduct is to be measured
by the happiness it produces.

Those who profess to be on the side of the angels are
usually too fearful of seeing *any* good in their opponents.
They refuse to admit that there is the slightest sense in
the arguments of the extreme left, and thus put weapons
into the hands of the foe. To say that there is no good in
hedonism is to be very, very hasty. Though it can hardly
be said that good is good for the reason that it produces
happiness, it can certainly be said that good does produce
happiness. And if you take into consideration the good of
everybody, counting not only present experience but all
possible future experience, and giving due precedence to
certain kinds of joy over certain more lowly (because less
intense, or more narrow, or more ephemeral) kinds of
enjoyment, you will have in hedonism a perfect measure
not only of morality but of holiness.

But what you make of it will obviously depend upon
what you believe, upon your philosophy in general, and
upon what kinds of joy you choose. So the theory is of
little moment in practice, apart from its valuable reminder
that conduct which is truly decent seldom produces (upon
others) much even immediate and almost no apparently
unnecessary anguish. Nor does hedonism furnish a theoretical basis for explaining the essential *nature* of an evil
act. Joy and sorrow, like pleasure and pain, are effects. So
the practical tendency of hedonism is, like that of epicure-
anism, to emphasize the immediate at the expense of the

remote, the common at the expense of the exquisite, the body at the expense of the higher faculties.

When I wrote *The Impuritans,* a far-Western critic professed to be much puzzled by my use of the words higher, lower, better, worse, etc., in a moral sense, and asked in effect, "What do you mean by higher?" I hate to think of him going all the rest of his life in doubt as to whether and why a good critic is better than a bad one, so I perhaps had better pause here to answer that, being an American, by higher and better I naturally mean *bigger.* I shouldn't always measure by a tape-measure, but I thought that Spencer had at least settled *this* point—that the higher forms of life are distinguished by the breadth of their outlook, the richness of their power of response, and the range of their appreciation.

But Spencer also attempted to settle another point—the question of how to get hedonistic principles adopted helpfully into the conduct of people in general. The kind of conduct he wished to see promoted was what is generally known as "Christian conduct." He called it "altruistic behavior." As this is founded upon a willingness to sacrifice one's personal and private worldly gains, and as he noted that it wasn't as common as might be, he turned the matter over to Lamarckian "Evolution."

An altruistic type, he thought, would in time be evolved —a kind of man who would get so much more pleasure out of seeing other people enjoy themselves than from having selfish pleasures of his own, that there would ensue "a competition in altruism." How this was to be brought about by the struggle for existence and the survival of the fittest he did not quite explain, for it is not altogether clear in what manner the selfish were to be killed off, unless the unselfish literally killed them with kindness. But anybody can see that if a competition in altruism had actually developed, say by the year 1914, Europe would have been startled by an exchange of diplomatic notes substantially to this effect:

Germany: If you want Alsace-Lorraine, please take it, for my sake.

France: My dear Fritz, I shouldn't think of doing such a thing. It would give you too much pain.

Germany: Not at all, my dear Frog. It would give me much pleasure to escape the pain which witnessing your pain has always given me.

France: In that case I might deprive myself of the pleasure of the pain of deprivation for the sake of the pleasure of witnessing your pleasure in the pain of depriving yourself of mine.

Germany: *Das nicht so gut ist.* If it is going to pain you to take it, I've got to keep it.

France: *Cela ne me convient du tout.* It would hurt your feelings too much.

Both: Let us set Alsace-Lorraine free!

There must have been a hitch somewhere. I fancy it arose from the difficulty of starting a competition in altruism in a world where opportunities for sacrifice are so abundant. The less vigorous altruists, though they might miss the big prizes, would still find food enough for their altruistic appetites, and would hardly fail to survive. Therefore the more altruistic germ-plasm would be continually contaminated with the less, thus preventing a growth from altruism to altruism, leaving us, in fact, in the very bog of unaltruistic altruism where the philosopher found us. He was looking for a change of heart, for that general unselfishness which would, everybody admits, make statesmanship so easy. But he sought it in a further development of that very selfishness which makes Utopia but a dream.

The modern Socialist does the same thing, though for altruism he says "enlightened selfishness." Where is the light coming from? Evidently from knowing upon which side your bread is buttered. But here we are confronted with the choice of butter. Can people be made to prefer the altruistic kind? Every man will choose according to his philosophy, which makes Philosophy rather important.

Philosophers, however, will not have it so. They are the only kind of people I know who keep insisting upon the unimportance of their calling. The notion that there is nothing in philosophy but the childish self-revelations of philosophers has taken such complete possession of the Rules Committee, that to treat thinking seriously when it is applied to serious things is considered off-side play.

"Place yourself . . . at the center of a man's philosophic vision," James admonishes the critic, "and you understand at once all the different things it makes him write or say." To be sure you do. But this, though it sounds well on account of that word "vision," is not putting yourself at the center of his system of thought and trying to see what he said and what it means; it is merely putting yourself in his place, and trying to discover how he came to say it. You must not criticize his books, you must content yourself with writing the dear boy's biography.

"Keep outside, use your post-mortem method, try to build the philosophy up out of the single phrases, taking first one and then another and seeking to make them fit, and of course you fail," James continues—as if there were no way of discussing a philosophy rather than a philosopher save by misreading its phrases, as if anybody could even be tempted to create inconsistency by twisting words when there are whole paragraphs, whole books of inconsistency at one's disposal!

But if you don't like flat contradictions and absurdities, it seems that "you crawl over the thing like a myopic ant over a building, tumbling into every microscopic crack or fissure," sometimes so microscopic that you could sling a cat through them, "finding nothing but inconsistencies, and never suspecting that a center exists." He is warning the Oxford students whom he is addressing what they may expect if ever the ant rather than the grasshopper should happen upon their works, and concludes: "I hope that some of the philosophers in this audience may occasionally have had something different from this intellectualistic type of criticism applied to their own works!" [1]

The more I study the brothers James, the more I prefer Henry, who could at least endure criticism without whimpering and calling it insectivorous. Criticism should not be intellectualistic, then, nor even intellectual. It should be—what? Gossipivorous. But I doubt if James would have liked even this kind, or have welcomed the critique which treated all he said as ravings not worth analysis but to be accounted for by those bad spells from which their author sometimes suffered.

What would you think of an architect, who should complain: "You have no business finding fault with my building, even if the roof does leak and the walls threaten to fall down upon the heads of passers-by. Talk about *me*. Tell folks that I am an upstanding man, even if I can't build. And think of the artistic emotion to be had from the sight of a dangerous building. All the better if it does

[1] *A Pluralistic Universe*, p. 263.

fall down and kill somebody. That will heighten the emotion."

James once wrote a book called *The Will to Believe,* but never once considered that there are two wills to believe, one anxious to find out and know the truth, the other no less anxious to avoid it. Truth is a nuisance. If it fails to obey our notions, that only shows you how the universe *ought* to have been made, what the center of the Creator's philosophic vision *ought* to have been! If it was otherwise, so much the worse for it. Let myopic ants bother with reality. Fifteen hundred and nine dogskins and a half are said to have been sufficient to furnish forth Gargantua with a pair of trousers. How many poor excuses are needed to cover the nakedness of a bad philosophy?

Bertrand Russell, speaking of ethics, says:[2] "I hardly think myself that it ought to be included in the domain of philosophy." But it is "traditionally a department of philosophy," and that is his "reason for discussing it"—his one noteworthy bow to tradition.

I think he was right in holding that it ought not to be included in the domain of philosophy, if it was his own philosophy that he had in mind. A world composed entirely of events that fail to eventuate, is clearly no place for ethics. When he comes to discuss them, he of course trespasses upon a more ancient philosophy.

By Ethics he understands the search for general principles, while specific rules for conduct he considers "the province of morals." Myself, I should like to find some words to distinguish between conduct regarded in the light of its worldly consequences, and conduct regarded in connection with a possible responsibility to a power other than

[2] *Philosophy,* p. 225.

that of the neighbors, for the confusion in current language is appalling. I might try to revive the words sin, virtue, righteousness and unrighteousness, but I fear to have them interpreted in the light of Jonathan Edwards. However I shall venture to suggest the term casuistry, in its ancient and uncorrupted sense, as the art of distinguishing between sin and saintliness. This leaves ethics to deal with the establishment of laws of a purely secular sort, most of them unwritten, whose essence is custom. It will soon appear how impossible it is to deal even with ethics without reference to a more profound philosophy.

It must be confessed, too, that these two fields frequently overlap. But the attempt to legislate against sin as such, to convert it into crime merely because it is regarded as sin, has always proved disastrous. For crime is suppressed by force, even when the force of policemen and prisons is not resorted to, and public opinion is left to work with its forceful instruments of ostracism, boycott, snubs, and tar-and-feathers.

Such punishment nearly always works injustice. It cannot be helped. Society, slow to take notice, lets many faults go unnoticed. Once aroused, it steps upon the victim with clumsy, crushing force. The only excuse for the resort to bodily coercion is necessity. The stigma "criminal" should not, I think, be attached to any conduct that does not trespass rather outrageously upon the liberty of others. Especially should one hesitate to invoke the formidable machinery of Law against trespassers who trespass only so to speak and by hurting other people's feelings when the feelings themselves are out looking for trouble. If any semblance of fairness is to be observed, the only crime to be drastically suppressed is crime that drastically imposes itself upon its victims. I have little sympathy with

the movement forcibly to save adults in the full possession of their senses and capacities from evils of which
they do not themselves complain. Must we always be
treated as children? How then shall we ever grow up?

Physically expressed force can never deal with sin, for
the very essence of moral obedience, using the word moral
to indicate a super-social obligation, is that it shall be
voluntary. If there indeed be no such thing as sin, as so
many wiseacres now contend, surely that is not a reason
for treating it as crime. And in dealing with crime we are
bound by common sense to limit ourselves to measures of
proved expediency. Just at present, we are confronted with
the spectacle of measures of demonstrated inexpediency
being supported, at least to a great extent, because they
are so worded as to condemn what some regard as sin—
supported, too, by many who do not believe that there is
such a thing as sin. I fail to see how even those who do
believe are excusable when they abandon their proper instrument, moral suasion, and resort to the bludgeon, which
is certainly one of the things which are Cæsar's. When the
churches resort to the criminal courts they cut a very poor
figure. No doubt church members are, as citizens, entitled
to their rights under just laws, but are they entitled to use
the Cross as a battering-ram against what they regard as
the ramparts of unrighteousness? Isn't this resisting evil
in that sense in which such resistance appears to have been
forbidden?

Russell, naturally enough, is somewhat bemused in his
discussion of right and wrong, for one reason because he
seems to have the idea that "rules" are not only "rigid,"
but so worded as to apply automatically to details—as if
an immoral act consisted in certain enjoined motions. But
the motions with which I stab my friend with a paper dag-

ger in a pantomime are the same as those with which I might stab my enemy in deadly earnest. I know of no code that does not distinguish between the two acts. Nor are such rules as have been actually enjoined either by ethical or religious teachers quite as "rigid" as he seems to think. The narrowest of them partake somewhat of the nature of general principles, and require some judgment before one can even determine whether they have been observed or not. He is over-anxious to discredit these "rules." But then nearly all modernists describe religion as something which, if true, everybody, from saints to monkeys, would unite in condemning. The object, no doubt, is to make everything that is even remotely associated with the word hateful to the public ear.

As to the ancient principle that evil arises from the conflict between man's will and God's, Russell will have none of it. And yet it may easily be maintained that this is the true principle, even without going beyond philosophy or learning for support upon a single dogma that is a dogma for revealed religion only. The moment we admit that man's force resides in his will, that the force of the universe resides in Another's will, one must admit sin as a contest between the wills of the human and divine. I doubt if the principle will be much felt in daily affairs if it goes no further than this abstract conception, but it will keep us at least from insane theories. Neither science nor philosophy can well demonstrate the specifically religious tenets of a creed, but they may be called upon to remove specious objections to religious belief.

The objection that the mere idea of a struggle between a human will and Omnipotence is a self-contradictory absurdity, though often voiced, is very specious. It assumed that Omnipotence could not create, that is, could not give to any other thing that sort of existence which would distinguish it in any way from Omnipotence itself —in other words, could not grant that independence of life which is commonly known as free-will. The objectors argue as if a will to be free must be free in all respects, so that its own action would be omnipotent, though it is obvious that a will is free if it be free to will, whatever may become of the effects it desires to produce upon other things.

If there be such a thing as a will—and there must be if there be conscious creatures—then the possibility of resistance, of sin, is a fact, to which the fact that we can't see how creation in this whole-hearted fashion was accomplished is totally irrelevant. This gives the capacity for sin to all conscious beings, even to animals. It must be possible to any being that does not react immediately and mechanically to its environment. But why should we not think that animals can do wrong in this natural sense? Any conscious preference given to the present over the future because one is now and the other is then, is a natural wrong. I know my dog is conscious of wrong when he permits the immediate pleasure of stealing a chop to outweigh the greater pain of estrangement and punishment to come. If he actually forgot himself, and reacted blindly to the chop stimulus, I should say that he was merely guilty of a crime. I frown and box his ears, not with a sense of moral superiority, but with the practical aim of giving him another stimulus that may be remembered next time not too late to restrain his jaws. If I were the dog's creator as well as master, the supreme being in the universe, and if Grits were capable of appreciating me as such, I should say that the dog's moral situation was upon all fours with ours.

Men who are but animals—I do not believe there are

such men, but let us suppose—would be morally upon all fours with the dog. All that could be expected would be canine conduct. It must always be ridiculous to hold any being morally responsible for insight which is not his— like saying that, morally, divorce is the same for one who believes as a Jew and for one who is brought up, for instance, as a Catholic. Yet this absurdity has often been asserted.

In this sense, I do not see how we can deny that morality is relative. This does not mean that there is no standard, for not even relativity can properly exist without a standard, but merely that we are limited by our limitations, and can never be sufficiently acquainted with the absolute standard to discover it in all its nature and apply it without error in all its fullness. But "relativity of morals" is an expression which is objectionable if it is used (as it commonly is) to suggest that there is no standard except an arbitrary one. If this be a quibble, then it is a quibble that distinguishes a church bell from belladonna. One aspires to be a stimulant, the other is a narcotic. Their effects are not the same.

Ethics, as the sum of worldly obligations, duties, and manners, is also of course relative in this sense. Also in a more superficial sense. Russell says rightly enough: "The rules of morals [meaning more particularly ethics as I have just used the word] differ according to the age, the race, and the creed of the community concerned. . . . Even within a homogeneous community differences of opinion arise. Should a man kill his wife's lover? The Church says no, the law says no, and common sense says no; yet many people would say yes, and juries often refuse to condemn. These doubtful cases arise when a moral rule is in process of changing." Quite. But I should hardly call such a community "homogeneous." Even its moral insight appears to be in process of changing.

Russell wishes to find something less subject to change than are the opinions of jurors.

"In a given community, an ethic which does not lead to the moral rules accepted by that community is considered immoral. . . . It does not, of course, follow that such an ethic is in fact false, since the moral rules of that community may be undesirable."

They already seem rather undesirable to me. But evidently there has been a moral invasion, or is a surviving morality, and this may be less "false" than the moral standard of a part of the community. There is, then, a "true" principle, somewhere. I am glad of that. But where and what is it?

"Some tribes of headhunters held that no man should marry until he can bring to the wedding the head of an enemy slain by himself. Those who question this moral rule are held to be encouraging license and lowering the standard of manliness. Nevertheless, we should not demand of an ethic that it should justify the moral rules of headhunters." Yes, but why shouldn't we?

"Perhaps the best way to approach the subject of ethics is to ask what is meant when a person says: 'You ought to do so-and-so,' or [the rarer case!] 'I *ought* to do so-and-so.'" This looks like a very good method of approach. But where do we arrive?

"Primarily, a sentence of this sort has an emotional content; it means 'this is the act towards which I feel the emotion of approval.' But we do not wish to leave the matter there." I should hope not. "We want to find something more objective and systematic and constant than a personal emotion. . . .

"The ethical teacher says: 'You ought to approve acts of such-and-such kinds.'" And such-and-such may possibly mean the kinds that bring home the head to the wedding. "He generally gives reasons for [his] view, and we have to examine what sorts of reasons are possible." Yes, yes. No one is disputing it.

"Historically, virtue consisted of obedience to authority, whether that of the gods, the government, or custom. In its more primitive form, the theory is unaware that different authorities take different views as to what constitutes virtue, and therefore universalises the practice of the community in which the theoriser lives. When other ages and nations are found to have different customs, these are condemned as abominations.

"The view we are now to examine is the theory that there are certain rules of conduct—e. g., the Decalogue—which determine virtue in all situations. The person who keeps all the rules is perfectly virtuous; the person who fails in this is wicked in proportion to the frequency of his failures." [3]

If Mr. Russell will read the Talmud he will discover that it is not the Decalogue that the Jews regard as a list of detailed acts, good and bad. And if he will turn to the thirteenth chapter of First Corinthians he will find that Saint Paul, certainly a fair representative of the Christian view, did not say that the person who kept all the rules was perfectly virtuous. He said (I quote the King James version, to which Mr. Russell need have no objection): "Though I speak with the tongues of men and of angels, and have not charity [i. e., love], I am become as sounding

[3] *Philosophy,* pp. 226–227.

brass, or a tinkling cymbal. And though I have all faith, so that I could remove mountains, and have not charity, I am nothing."

Evidently, though historically virtue was unquestionably thought to be shown by obedience, it was early acknowledged that, aside from the importance of the particular authority, it was of some moment whether the obedience was voluntary, and if voluntary, what was its motive. Russell is merely trying to discredit earlier times, for he admits the importance of freedom and of motive when he says: [4] "The good life is one inspired by love and guided by knowledge." It is knowledge, however, which we are more particularly trying to get at. Is this a sample?

"One can argue [against the utilitarian theory] that there is more happiness to be derived from love than from hate"; also that "people cannot love to order." [5] This is denying that there is such a thing as a will to love. As to the pleasures of hate, he says that he means it is not always true "in *an individual* case, that love brings more happiness than hate." So he thinks that "ethics is mainly moral," a matter of rules rather than principles.

"The attitude of a neutral authority," he goes on [6] would be this: "Men desire all sorts of things, and in themselves all desires, taken singly, are on a level [*sic!*], i. e., there is no reason to prefer the satisfaction of one to the satisfaction of another." Evidently he means that love is as good as hate, so far as the lover or hater alone is concerned! "But when we consider not a single desire but a group of desires, there is this difference, that sometimes all the desires in a group can be satisfied, whereas in other cases the satisfaction of some of the desires in the group

[4] *Ibid.*, p. 235.
[5] *Ibid.*, p. 232.
[6] *Ibid.*, p. 233.

is incompatible with that of others." This seems to refer to a group of desires in a single person, but if it does it is hardly true, for hate is no more incompatible with love than is love with hate.

Be this as it may, " 'good,' " he says, "comes to apply to things desired by the whole social group," a group of persons, not of desires, being obviously in his mind. So he concludes: "It is evident, therefore, that there can be more good in a world where the desires of different individuals harmonise than in one where they conflict. The supreme moral rule should therefore be: Act so as to produce harmonious rather than discordant desires." [7]

This is what is sometimes called "the rule of coöperation. In practice, I fear it would be difficult to distinguish it from the rule of the stronger. Coöperation would become sauce for the geese, and hedonism for the ganders. The outstanding bit of truth in Russell's ethics is that the moral nature of conduct may be judged by the non-conflicting character of the desires from which it springs. But is this not because of the nature of love itself, which (not to confuse it with passion) is unselfish, and seeks for none of those things of which there can possibly be a scarcity?

The approval of the community has nothing to do with it. In a community of children whom you loved you would not ethically coöperate with their desires to see a match applied to a barrel of gunpowder. And if it be argued that the childrens' desires, though prompted solely by the love of fireworks, are not of the sort not to conflict in the long run with their other desires, such as that of continuing to have arms and legs, I shall argue back that precisely this situation occurs in adult society. So the moral man or

[7] *Ibid.*, p. 235.

woman is compelled not to further what he believes to be mad practices, no matter how widely approved, which, according to Russell, makes the moral become the immoral by the very force of their morality.

What he really does is to cast his vote for standardization, for the establishment of a moral code by counting noses, and perhaps dollars. He certainly offers little aid and comfort to those who find themselves in a community whose aims and intelligence are different from their own. And yet nobody, least of all Mr. Russell, would admit to friendship and upon terms of equality anybody who was guided solely by the public opinion of the moment; who was entirely "moral" in a "social" sense, never dared to stand alone, but was always considering the consequences and perpetually offering himself up at auction to the highest bidder. In all ages human admiration has eventually clung only to those who, at the great pinch, were capable of saying, "Damn the consequences!"—meaning precisely the immediate social consequences.

Whether the actual desires of a group are going, in the long run, to harmonize in the production of their greater happiness, cannot be told by the ballot box. The ethic which Russell professes therefore breaks down even as a theory. Morality does not tend to get itself identified with popularity. It is much more likely to get identified with unpopularity—though if it could rid itself of imitation, and of self-righeousness, it would doubtless be much less unpopular than it is. Prophets, though they were obnoxious to rulers, have been known to be welcomed by their fellow sufferers from the principle of coöperation—people who were on the wrong side of the deal and in contact, perhaps, with the wrong coöperation.

In Walter Lippmann's widely commended *Preface to Morals*,[8] it is said that "the long record of clerical opposition to certain kinds of scientific inquiry has a touch of dignity when it is realized that at the core of that opposition there is a very profound understanding of the religious needs of ordinary men. For once you weaken the belief that the central facts taught by the churches are facts in the most literal and absolute sense, the disintegration of the popular religions begins."[9]

Clerical opposition has not always been wise. But most of it has been directed against the sort of science which sought to show that "the facts taught by the churches" are not true in any sense except in the sense of being untrue. And now philosophers, pretending to speak for science, and maintaining their own opposition to the clerical opposition, are telling us that their own facts, which we must continue to accept as refutation of the more clerical facts, are not true, either. This is scarcely dignified.

But Mr. Lippmann seems to indicate that the clerics remain dignified, and even justified, when they oppose the search for facts because they wish to preserve the currency of certain lies which, though they do not believe them themselves, they consider good for the people at large. For my part, I should say that when clerical opposition begins to oppose men's making use of such light as there is, it ceases to be dignified.

Of course, no two men upon different intellectual levels, *can* interpret the same fact in quite the same way. One will necessarily see more in it than will the other. And this might lead one of them to object to the promulgation—not of facts, but of fables—such as the fable that says that a

[8] New York: The Macmillan Co., 1929.
[9] P. 35.

narrow view of anything shows all there is to it. Just possibly, even a cleric—though the word includes all sorts and conditions of men—may sometimes be merely suggesting that the less informed should not make terms for the more, or the half-informed play havoc with the uninformed.

In any case, is there after all such a wide difference between "the religious needs of ordinary men," and the religious needs of the man extraordinary? None of us are so extraordinary as not to be in the same boat with our fellows when it comes to the real ocean of mystery. And what everybody needs, even in the sea of ethics, is something that will keep him from drowning in the multitude of waves, something to appeal to *from* Cæsar.

We need Cæsar to regulate the competitive struggle for things of which there are not enough to go around. But we need something else to lessen the struggle, say by turning the attention towards prizes of a different sort. It is amusing to realize that we could all be rich, even in worldly goods, if we were not so anxious for them and were willing to produce them for others without the near-certainty of getting a slice for ourselves. It is nearly impossible, for example, for a religious community not to grow rich. Wealth is the curse of the pious, and fairly dogs their steps, while avarice spills the beans by grabbing too hard at the bag.

But what does Mr. Lippmann mean when he says that "fundamentally the great churches are secular institutions; they are governments preoccupied inevitably with the regulation of the unregenerate appetites of mankind?" [10] Can a Church be called great, or be called a church at all, that is occupied *fundamentally* with the regulation of its secu-

10 P. 201.

lar affairs, or even with the regulation of the unregenerate appetites of mankind? To regulate an appetite while leaving it unregenerate, is to regulate it by force. Surely the least of *churches* must be trying to regenerate, to change these appetites, to supplant them with new and different ones. An advertising agency could go as far as that.

And then Mr. Lippmann tells us that in the scriptures of "the great ecclesiastical establishments . . . there is to be found the teaching that true salvation depends upon the internal reform of desire." He seems internally to have reformed his opinions. Or does he mean that these great establishments have fallen out with their own scriptures? This seems to be it, for on page 216 he discovers that "the God of the Book of Job" does not minister to human desires, and that "the story of Job is really the story of man's renunciation of the belief in such a God. . . . The God whose ways Job finally acknowledges" being "no longer a projection of Job's desires." No longer a projection of Job's original "unregenerate" desires, especially. But Mr. Lippmann will have it that Job's finally-discovered God "seems to be consistent with the orthodox popular religion," but is "really wholly inconsistent with the inwardness of the popular religion."

I fancy this is true of much "popular religion," but it seems to restore their fundamentally non-secular character to the "great" and presumably "orthodox" churches, to say nothing of the great ecclesiastical establishments. I think that Mr. Lippmann knows more about Job than about the Church, for it is a cold historical fact that the Church has always taught what he has discovered to be the teaching of the Book of Job, and so have most of the Churches, though God alone knows what an occasional ecclesiastic may or may not have taught.

Mr. Lippmann laments—as who does not?—that "the symbols of religion are choked with the debris of dead notions in which men are unable to believe and unwilling to disbelieve." [11] But why add to the debris this other dead notion, at the bottom of the same page: "Since they [men] are unable to find a principle of order in the authority of a will outside themselves, there is no place they can find it except in an ideal of the human personality"? An ideal of the human personality is a very beautiful thing, but if those who are unable to find a principle of order or the authority of a will outside of either that or themselves, will kindly try jumping out of a second-story window, say, with the will not to get hurt, they will find what they are supposed to be looking for without even going to church— unless their wills be exceptionally powerful.

Will Durant, in his *Mansions of Philosophy*,[12] also has his word to say—though whether it prefaces morals I am not so sure. "We should realize," he tells us,[13] "that the coöperation in which morality consists arises less from the growth of the soul than from the widening necessities of economic life. . . . Morality spreads [he means that a given code of custom spreads] as economic and social units increase; the whole with which the part must co-operate to be saved becomes greater as the world is woven into ever larger units by rails and wires and ships and the invisible bonds of the air. Once trade and common interest merged tribes into nations, and tribal morality degenerated into the last refuge of a scoundrel. Slowly trade and common interest merge nations into vast national groups, and provide the basis for an international morality. Soon all the world will agree that patriotism is not enough."

[11] P. 328.
[12] New York: Simon and Schuster, 1928.
[13] P. 140.

There are those ready to admit that patriotism is not enough now, and some who even go so far as to think that coöperation with rails, wires, ships, and airplanes is not enough. What is enough? According to Durant, it will be enough to coöperate with the League of Nations, when that organization finally has teeth in it. My one consolation is the thought that when the economic and social unit with which the part must coöperate has become as large as that, the last refuge of scoundrels will have correspondingly widened, so as to include even patriotism. For any but scoundrels, it looks as if morality were going to become hard sledding—especially if there is to be no coöperation with a yet-larger Whole, say one large enough to have some private relationship with individuals as such. The League secretariat is almost certain to be too busy remaining in office to give much attention to minor complaints.

"No sooner do we reach our definition of morality as the coöperation of the part with the whole," Durant continues, "than a hundred new questions appear. With what group shall we coöperate—with the family, or the state, or humanity, or life?"

This evidently refers to the interval during which we have reached our definition but not our goal. For there are still more or less independent groups. Prudence seems to suggest that one would do well to coöperate with the most influential. Does this mean "life"? But it can't be life in anything but a social, or economic sense, since it arrives by way of the amalgamation of the rail, wire, ship and airplane interests. But however imprudent—

"When a man turns forty his great temptation is to conceive morality as solely devotion to his family. Not that he quite lives up to his conception; if he did, perhaps (as Confucius thought) no other morality would be required. If the state has grown like a leviathan, and has absorbed one parental right and function after another [I do hope it will never absorb them *all!*] it is not merely because our economic life has developed complex interrelations and contradictions which demand at the center of the community a coördinating and adjudicating authority; it is also because the individualism of industry has disintegrated the parental authority." [14]

I wonder if it is not the individualism of industrialists rather than of industry which has done all this? But I gather that they, or it, have brought or are bringing about a World State, wherein individualism, among employees at least, will be confined to scoundrels—in so far as they are unconfined. Somehow, I don't so much blame the man turned forty—with whom, perhaps, I have a fellow feeling. I am rather sorry that he has not lived up to his conception, which would have given the "individualism of industry" a harder row to hoe.

But it is hardly fair for *pater familias* to lay all the blame upon rails and wires. His own unwise wielding of parental authority may have had something to do with the situation. In any case, our coöperative morality is bringing us, some say, to the point where the average young man finds it difficult to get married at all or to lay the foundations of parental authority with as good a prospect as might be desired of having sufficient funds for the maintenance of decent surroundings and the ancient modicum of self-respect. Again the fault may be partly the young man's, and the young woman's, who play industry's game a bit too heartily by classing among the things which they cannot do without every bauble that machinery can be made to make. Isn't the recent plea that women must wear

[14] *Loc. cit.*

long skirts so as to support the cloth manufacturers, going a little too far? Some folks, if it be only the women themselves, have got to support the women.

It must be said on behalf of millionaires, that they have done their best to keep the rougher edges of the growing rail-and-wire World State from encroaching upon their own families. But who shall say what will happen if our ever-rising standard of living shall become absolutely risen and standardized, with all things so cheap that only scoundrels can afford to buy them?

This "morality" trying to stand alone seems to fare much like matter trying to stand alone. As Durant himself asks, "Must our moral code resolve itself into loyalty to politicians? That is the answer which the politicians give." And he thinks it "is not quite without reason," at the present moment. "For until an international order is a reality, and humanity organized to use and protect the allegiance of the individual, an ideally perfect morality—a coöperation of the part with the completest whole—will be but a counsel of perfection." [15]

Nobody can claim that Durant does not go the whole hog. All hail the ideally perfect League, ideally organized to protect the individual not only from other individuals but from the League. Evidently there are going to be no politicians in it, and so no man of forty will ever again be tempted to put the welfare of his family above the economic life of the world. Possibly he won't be permitted to.

I have no quarrel with the State, as such, or with the present League of Nations. But before it assumes the direction of all human activities, I should like to feel certain that I would be comfortably near the top, just in case the ethics of coöperative morality should become the ethics of

[15] P. 141.

tyranny. One great difference between a Church and a State is that one doesn't have to join the Church, however universal. To recognize its authority is a voluntary act. But it might be difficult not to join a World State.

For such an organization to be safe for anybody but the favored few, it rather looks as if there would have to be *something,* some other widely recognized standard, as little interested in rails, wires and ships as possible, to which one could cry for the preservation of a certain individual autonomy. If this Something went so far as to admit the peculiar importance of the human soul, no great harm might be done—in the long run. For not even a territorially limited State can show at its best if its citizens are one hundred per cent. citizen, and not any per cent. men and women. A collection of rabbits and rabbit warders seems hardly a counsel of perfection for any world but that of William Clissold, and I believe that an attempt to reduce men to rabbits in toto would meet with considerable resistance. With no restraining influence higher than the collective will, which of course would be manipulated by interested parties, nullified by division and by the pitting of itself against itself; with no shadow above the heads of dictators save that cast by the necessity for diplomacy, we should all find ourselves fighting like tarantulas in a bottle, or—the worse case—reduced to rubber stamps. Durant's touchingly naïve faith in the perfectibility of Big Business moves me to tears. So I turn back to Dewey.

"Convert the objects of knowledge into real things by themselves," he says,[16] "and individuals become anomalous or unreal; they are not individualized for science or law. The difficulty under which morals labor in this is evident. They can be 'saved' only by the supposition of another kind

[16] *Experience and Nature,* pp. 147–156.

of Being from that with which natural sciences are concerned."

I started to quote this because it sounded relevant, but now that I read it over for the thousandth time I can't quite see what it is relevant to. The idea that neither morals nor individuals can be "saved" except by something other than that which is the subject-matter of the natural sciences, is promising. But you will note that this Being is to be supposed only in case we convert the objects of knowledge into real things by themselves. And that is going to make individuals either anomalous (that is, abnormal) or else unreal, or both. Who wants unreal individuals saved? And morals, no matter what the difficulty under which it labors, ought to have some dealings even with those who are only normal.

Also I find myself laboring under considerable difficulty trying to suppose that the objects of knowledge are either unreal or are real *in themselves*. A thing real in itself, self-exists. Is this sheet of paper self-existent, or isn't it an object of knowledge? Or is it my knowledge that is self-existent, and has no need of paper? Or is Dewey here hinting at that great truth that we see through paper to its self-existent cause? But how can we if we are unreal? Evidently he means that we are self-existent, and our own objects. But this "ethic" is almost too anomalous. Let's try another spot, say the beginning of Chapter Ten:

"Recent philosophy has witnessed the rise of a theory of value. Value as it usually figures in this discussion, marks a desperate attempt to combine the obvious empirical fact that objects are qualified with good and bad, with philosophic deliverances which, in isolating man from nature, qualitative individualities from the world, render this fact anomalous."

I think this means—though I am quite willing to stand corrected if anybody now living will come forward and say that he understands it or can say definitely in what language this whole Chapter Ten is written—I think it means that people have long since noticed that some objects are good, in the sense of pleasant to the noticer, and that some are bad, in the sense of unpleasant. But a desperate attempt has been going on to reconcile this fact with a philosophy (not Dewey's) which isolates man from nature and makes him of a quality different from that of the world.

"The philosopher," and now I suppose he means the "real" philosopher, "erects a 'real of values' in which to place all the precious things that are extruded from natural existence because of isolations artificially introduced." I hope I don't extrude if I suggest that this means that the real philosopher (real in Dewey's estimation) tries to find a place in his system of thought for the precious things, or feelings, which false philosophers have tried to keep away from man by putting the things in one box, called nature, and putting man in another box, called something else. And Dewey proposed to put everything, man and values, into one box.

If there were any widespread belief to the effect that *no* part of man was in nature, this procedure of Dewey's might have yet more value than it already promises to have. But what has this to do with morals? You know very well what it has to do with morals. It means that man's sense of the value of an object *is* the value of an object, and that his sense of the object in any sense is also the object, and vice versa, because there isn't any object. It hardly seems necessary to "save" morals now.

"Poignancy, humor, zest, tragedy, beauty, prosperity

and bafflement," he adds, "although rejected from a nature which is identified with mechanical structure, remain just what they empirically are," that is, just what they seem to be, "and demand recognition."

I, myself, never intend to deny them recognition. My feelings of poignancy, humor, zest, tragedy, beauty, prosperity and bafflement, though bafflement is now in the ascendent, do not often remain just as they are (not for long), yet they are what they are while they are it, and as feelings I willingly reject them from a nature that is identified with mechanical structure. But I don't see how I become anomalous or unreal or put morals into any special difficulty if I go on and say that my poignancy, and all the rest, are not only real feelings but are really provoked by something.

"When we return to the conception of potentiality and actuality, contingency and regularity, qualitatively diverse individuality, with which Greek thought operated, we find no room for a theory of values separate from a theory of nature." O wad some fey the giftie gie me to talk like this! But I gather that Dewey is, among other things, laboring to deny that good and bad are objects having a life of their own and floating about apart from all "existences." I move it be conceded. But if he means that the Greeks found no room for theory of value apart from mundane nature, he still has a Greek author or two to read.

"We must surrender," he says,[17] "the identification of natural ends with good and perfection; recognizing that a natural end, apart from endeavor expressing choice, has no intrinsic eulogistic quality, but is the boundary which writes 'Finis' to a chapter of history enscribed by a mov-

[17] P. 395.

ing system of energies. Failure by exhaustion as well as
by triumph may constitute an end; death, ignorance, as
well as life, are finalities. . . . [We] must abandon the
notion of a predetermined limited number of ends inherently arranged in an order of increasing comprehensive-
ness and finality."

This, in spite of its warning not to identify natural ends
with perfection, has the air of intending to signify that
one end is as good as another, since all ends are equally
final—life, apparently, being final even before it comes to
its end. Only ends having endeavor expressing choice are
praiseworthy, but not even these are more comprehensively
final than natural ends. Less so, it would seem, for it is the
end expressing no choice which brings them to an end and
writes the final Finis—though the use of the word "end"
now in the sense of a full stop and now in the sense of
"aim," renders this conclusion endlessly doubtful. My
opinion of this passage has no intrinsic eulogistic quality.
But if you think you have not had good "value," read his
entire chapter and see if you can find more.

Professor Alexander [18] lays the burden of morality
where it belongs, on the will, but seems to limit the words
"good" and "evil" to acts where the will actually succeeds
in bringing some outer event to pass; or as he puts it,
"bring(s) into existence certain external relations among
real things corresponding to the idea first entertained."
This would make it impossible to sin alone and when not
moving. But as probably no act of the will is followed by
no physical result whatever, if even but an extra heart-
beat, let us pass to the test that he applies.

"The reality which we produce," he maintains, "is good

[18] *Vide, Space, Time and Deity,* Vol. II, Bk., III, Ch. IX, C., p.
273, *et seq.*

in so far as it satisfies coherently the persons who bring it about. . . . The facts we seek to bring about are, so far as their good is concerned, determined by how far they satisfy persons and are approved by them."

The word "coherently" evidently applies to persons who cohere in society, especially as he immediately adds: "All action is response to the environment, but one part and the more important part of our environment in moral, that is social, action, is our fellow-men. For not only do we take account of their approbations as we do in the prosecution of knowledge, but they are themselves the objects of our appetites, as food and drink are. . . . Morality is the adaptation of human action to the environment under the condition set by the environment. . . . Morality arises out of our human affections and desires which we seek to satisfy. Some of them are self-regarding, others are natural affections for others. . . . The good act, approved as pleasing to the collective wills and not merely the individual's own will, may vary according to the nature of the individual and the place he holds in the society. . . . Just as truth resides in the union of reality with the minds which possess truth, so goodness resides not in the bare satisfaction of appetites alone nor in the will alone, but in the union of satisfying objects with the wills which sustain them." [19]

Does this not make the good man merely the popular man, the man approved by his neighbors? Such a man would certainly never get into any trouble—unless there happened to be a change of administration. He would always be doing what the majority in influence thought he ought and willed that he should do. And if he remained a

[19] *Ibid.,* Vol. II, pp. 274-277.

good man through all the vicissitudes of shifting fashion and policy, he would still be the Vicar of Bray.

Alexander tries to confuse the issue by suggesting that this procedure is similar to that whereby we acquire knowledge when we go to school and listen to teachers. This calls up the picture of a member of a religious community, acquiring moral virtues through obedience to his superiors. But if he be moral he doesn't do it by submitting to the "will" of his superiors or of the community. He can become moral only through his own convictions.

"Badness, or moral evil," says Alexander,[20] "is the same reality with which morality is concerned, handled amiss."

And again: "Much suffering and heart-burning may be endured in the social adjustment of claims," and in the "exaltation of what is approved of them into rights, till the individual has learnt the difficult lesson of finding more pleasure in following the right [that is, the popular in his vicinity] than he loses from the sacrifice of his desires. There are even claims which must be called natural, though there can be no natural rights. Such are the elementary claims for freedom and life, which no society can refuse to turn into rights without compassing its own destruction." [21]

But of this the powers that be must judge, and there have been times in history when they did not hesitate thus to encompass their own destruction. Any and all tyrannies should be able to get through this needle's eye of Alexander's, at least if they be republican in form. What becomes of the individual, without rights either natural or supernatural, who finds himself unable to take pleasure in

[20] *Ibid.*, Vol. II, p. 280.
[21] Vol. II, p. 283.

the collective right—or is it rite?—of headhunting? Obviously he becomes immoral, and so a fit subject to furnish the next head without hunting for it. Or else he behaves in a way that looks moral, and hunts heads against the grain.

"The terms of moral disapproval indicate the process by which the unsocial type," the type that does not really like headhunting and simply will not hunt, "is discarded in human life. The elimination which in nature is accomplished by death is here accomplished not by death, except in extreme cases where the deviation from the type is too great for mercy, but by the sentence of exclusion, which leaves room for the individual censured to return to the type on condition of altering his character if he can." [22] Or at least pretending to, if he can't. Otherwise, beware of deviating from the type which quantity production in your neighborhood sets up for worship! This makes all heroism for principle not only a crime, but immoral. I wonder if Professor Alexander, in so far as morality is concerned, has not handled his subject somewhat amiss? Let us ask Professor Whitehead.

I regret that I am unaware of Whitehead's complete ethical system. But I recall one bit from his *Science and the Modern World* in a chapter dealing with science and religion:

"Evil is the brute motive force of fragmentary purpose, disregarding the eternal vision. Evil is over-ruling, retarding, hurting. The power of God is the worship He inspires. That religion is strong which in its ritual and its modes of thought evokes an apprehension of the commanding vision. The worship of God is not a rule of safety— it is an adventure of the spirit, a flight after the unattain-

[22] *Ibid.,* Vol. II, p. 285.

able. The death of religion comes with the repression of the high hope of adventure."

How beautiful is the sound of common sense! And though the power of God as creator can hardly be limited by the worship He inspires, it is certainly true that here is a field never to be attained in the sense of exhausted, and true also that the worship of God is not a rule of worldy safety, any more than morality consists in at all times angling for the immediate approbation of our fellow men.

APPENDIX

A. VEDANTIC PHILOSOPHY

It is impossible to think of the great will of the universe as evil, for we have but to surrender to it and its ways become our ways and set us free. But where I quarrel with the great philosophies of India is at the point where they affirm that life is a pouring forth from Brahma—a stream, which, moving in a circle, returns eventually to its source. What a poor reward for love, to extinguish it at the end by giving it its full desire, like a moth plunging into a candle! Think of all this fearful to-do of earthly existence, merely for the sake of getting back to where we started from! If creation was for the sake of something that could not be had without it, why must it be undone and brought to an end? Or forced to repeat itself? If it could begin, can it not continue?

Besides, this Vedantic arithmetic strikes me as being of the worst. Perfect surrender implies perfect comprehension. But the difference between the finite and the infinite is itself infinite. To say then that we progress along an infinite way and yet arrive at its termination in Oneness is a contradiction in terms. I do not pretend to be in Brahma's confidence, but I myself am able to think of something much better than this—say an endless inventiveness which always leaves me something yet to discover. This proves nothing? Perhaps not. But I very much doubt if my own conception of the highest possible good exceeds the Creator's.

But to be fair to Vedantic philosophy one should read it in its great originals, and not as diluted and belittled by modern Swamis. For example, the *Bhagavad Gita* says:

"And he who by the power of Spiritual Wisdom doth perceive this difference between the Soul and the Material and Personal Self; between the Soul and Nature and Nature's Principles—between the knower and the Known —verily he perceiveth the liberation of the Soul from the illusion of Matter and Personality, and he passeth to the Spiritual Consciousness, in which all is seen as One Reality, without illusion or Error."

How refreshing this, compared with Vivekananda. Here the power of spiritual wisdom manifests itself by perceiving a *difference*. And taking the passage as a whole it may easily be interpreted as concluding with a vague yearning towards union, rather than a positive affirmation that any such consummation has actually been attained.

In Maeterlinck one sees a failure to understand his sources similar to that shown by Vivekananda and James. What troubled Maeterlinck was the difficulty of crossing the gulf between the absolute and the relative, once he had, in fancy, put himself in the Absolute's place. The fathers of the Church whom he quotes were cautious enough never to put themselves there, knowing that it was impossible to do so in fact; and that difficulties created by fancies are pseudo difficulties, that there need be no logical jump from the absolute to the relative because the mind-made absolute is a mere fiction.

Says Maeterlinck, in the work already quoted: "According to Scotus Erigenus, the great theologian of the ninth century, who reproduces the doctrine of the Areopagite, God is Being without predicates, above all the categories—that is to say, Nothingness; that is to say, the incomprehensible essence of the Universe."

We have already seen that the only sense in which Nothingness can here be used is in the sense of Incomprehensible. But Maeterlinck believes he has the Church on the run, and means to pursue his advantage ruthlessly until he has convicted Saint Peter himself of heresy. "This negative theology," he goes on, "has never been condemned by the Church, and is found even in Bossuet, the most affirmative, rigid and orthodox theologian the world has seen. . . . 'The whole vision of faith,' he [Bossuet] tells us, 'seems to be reduced to seeing plainly that we see nothing. And when we say that the soul sees God by faith, this is merely saying that it does not see Him,'— words which take hands across fifty centuries with those of the great doctrines of India, and notably with the phrases of the Sama-Veda."

But why *should* the Church, or the world either for that matter, condemn a negative theology which is negative merely in its refusal to reduce the Incomprehensible to a spurious Comprehensible? And why not shake (or even take) hands with the great doctrines of India in so far as they were great?

What Maeterlinck is really trying to get at it would be difficult to say. Perhaps it is "nothing" in the vulgar sense of the word.

B. FECHNER

As professor of physics at the University of Leipzig, Fechner won considerable fame from his experiments with galvanism and from his observations on the after-images in the retina. But he had another side, which he expressed in a series of poems, essays, half-humorous and half-philosophic romances, including a "vision" of the supposed inner life of plants, of a weird interest as fiction. His breakdown seems to have been due to eye-strain, general over-work, poverty, and fundamental lack of balance. His physical health took a turn for the better, and he deemed his recovery miraculous, and devoted the remainder of his life to scattering "daylight," as he termed it. Here is a specimen passage of James's own approving paraphrase of Fechner's *Zend-Avesta:*

"The entire earth on which we live must have its own collective consciousness. So must each sun, moon, and planet; so must the whole solar system have its own wider consciousness, in which the consciousness of our earth plays one part. So has the entire starry system. And if that starry system be not the sum of all that is, materially considered, then that whole system along with whatever else may be, is the body of that absolutely totalized conciousness of the universe to which men give the name of God. We must suppose that my consciousness of myself and yours of yourself, although in their immediacy they keep separate and know nothing of each other, are yet known and used together in a higher consciousness, that of the human race, say, into which they enter as constituent parts. Similarly, the whole human and animal kingdoms come together as conditions of a consciousness of still wider scope. This combines in the soul of the earth with the consciousness of the vegetable kingdom, which in turn contributes its share of experience to that of the whole solar system, and so on from synthesis to synthesis and height to height.

"If the heavens really are the home of angels, the heavenly bodies must be those very angels. The earth is our great common guardian angel, who watches over all our interests combined. The more inclusive forms of consciousness are in part constituted by the more limited forms"—the other part apparently coming from the pooling of the general information. Our mind "is not the bare sum of our sights plus our sounds plus our pains. In adding these terms together [it] finds relations among them and weaves them into schemes and forms and objects of which no one sense in its separate estate knows anything. So the earth-soul traces relations between the contents of my mind and the contents of yours, of which neither of our separate minds is conscious." [1]

Evidently the Fechner idea, which James momentarily adopted, was that the mere getting together of matter produces a mind capable of arriving at information unknown to any of its constituent parts. This matter creates something beyond itself. In a less crude form, the notion clung to James to the end. Its vice is that it tries to hide the miraculous by making it an attribute of the lesser.

C. TIME'S CHANGES

Bergson, however, takes the sting out of feeling by making it theoretically impossible. He finds that existing consists in passing from one psychic "state" to another, and that for the convenience we think of each state "as if it formed a block," though actually the states themselves are "undergoing change" every moment, and run together. "If a mental state ceased to vary, its duration would cease to flow."

This, of course, is in flat defiance of all that we *know* of either psychology or physics. Movement in nature is discontinuous and always proceeds in such a way as to manifest a minimum quantity of energy known to mathematicians as h, or some multiple thereof. A certain amount has to accumulate of anything, even of time, before we are capable of noticing it.

That *something* is continuous, I do not doubt; and that

[1] See *A Pluralistic Universe*, edition cited p. 168 *et seq*. I have condensed the original text.

we have something of it within us is evident, or we could not connect one moment with another. Bergson would have it that this continuity is continuous change, though how he knows it, or what he stands on to see it change, if all is change, he deigns not to explain. To speak of change as something *per se* is, in fact, meaningless, like Relativity without a body of reference. When we speak even of the Changeless we mean only changeless in relation with that which we call moving. As usual, however, Bergson is attempting to do away with the imponderable by pretending to ponder upon it.

"The truth is," he declares, "that we change without ceasing, and that the state [which we call a state of consciousness] itself is nothing but change. . . . The transition is continuous. . . . Each [incident] is borne by the fluid mass of our whole psychical existence. . . . But as our attention has distinguished and separated them [the incidents] artificially, it is obliged next to reunite them by an artificial bond. It imagines, therefore, a formless ego, indifferent and unchangeable, on which it threads the psychic states. . . . Instead of a flux of fleeting shades merging into each other, it perceives distinct and, so to speak, solid colors set side by side like the beads of a necklace. . . . As a matter of fact, this substratum [or thread] has no reality. . . . If our existence were composed of separate states with an impassive ego to unite them, for us there would be no duration. For an ego which does not change, does not endure; and a psychic state which remains the same so long as it is not replaced by the following state, does not endure either. . . . As regards the psychical life unfolding beneath the symbols which conceal it, we readily perceive that time is just the stuff it is made of." [1]

[1] See *Creative Evolution*, Ch. I, p. 1-2.

The "truth," or indeed the meaning of any sort to be discovered in this rhetoric, is small. He seems to assume first that there is a change going on outside of our consciousness, that we become conscious of its result at intervals, and then try to reconstruct the unobserved intervening process in imagination. So far, so good. It must be admitted that we do not ordinarily perceive that minimum quantity, known as h, which the new physics calls *quantum*. But then he claims that this intervening process which we imagine, is *merely* imaginary, and has no reality outside of us. He is drifting towards Solipsism. At the same time he is trying to penetrate within the unperceived depths of quantum, and catch creation at work—clearly an impossible job. We only catch creation after it has created and in what it has created. On top of this, he must seek to destroy the "impassive" ego, or indeed any and all kinds of ego, and make change out of change, ringing endless changes upon change. So Time is made of change, and change is made of time. And the things that change, if any such there be, are made of their changing. This may be music, but it is the music of muddle rather than of the spheres.

D. SEEING IS BELIEVING

Faith has been called the evidence of things not seen, and I am using it in this sense of *evidence* when I say it cannot be deceived. When it is suggested that the evidence is not seen, that merely means that in the case of religious faith, the evidence is not seen with the bodily eye, or any of the physical, or "natural," senses, but only with another sort of eye, especially opened. If there be such an eye, if one has such an eye, it is absurd to say that it does not see what it does see, for it is impossible that we do not have those experiences which we do have. The subject of Divine Revelation lies far beyond the scope of this book, which limits itself to the bare mention of the nature of the philosophy involved in the doctrine. I wish only to remark that deceived faith is a contradiction in terms, like false evidence.

Testimony may be false, not evidence. For testimony is a conclusion arrived at, perhaps honestly, from evidence which was incomplete, or from other testimony, which may be false. Or it may be in itself wilfully false. If we listen to a witness, we may doubt that he is telling the truth, or even what he believes to be the truth, but we cannot doubt that we are having that experience which we call listening to a witness. This is evidence. The rest is testimony. The expression "walking by faith and not by sight" I take to mean walking by the light of incomplete experience, or incomplete faith, or incomplete good-will, which darkens such light as we have. Walking by sight would then mean that the incompleteness had been done away with.

The sources of error are always the same, no matter what question is involved. It is customary to say that the bodily eye can be deceived, but if we mean the sense of bodily vision this is not strictly the case. If we see a witness on the stand, it may possibly be a wax dummy concealing a phonograph, but this is merely saying that a certain experience, which we call the experience of seeing a witness, which we expect to be followed by another experience, which we will call touching a human being, or talking to him, or observing him stand down, may in reality be followed by experiences which we call touching wax and seeing a piece of machinery. It is always the

experiences that we have not had which are in doubt.
We mistake a distant candle for a star only when for
the moment it fulfills all the functions of a star. Our error
comes from assuming that it will continue to fulfill all the
functions of a star. Some would say that our reasoning is
wrong, but the fault is not in the reasoning but in the lack
of evidence upon which it is based. Now we walk towards
the supposed star, and it begins to show a wick, a flickering
flame, a stick of grease stuck into a candle-stick—very
unstarlike behavior. So we conclude, from this new evidence, that it is a penny dip. Again it is not the reasoning
that is at fault, but the evidence. For as we walk forward
still farther we have that sensation which we call encountering a cold flat surface—quite new evidence. We are immediately constrained to say that we are looking at the
reflection of a candle in a mirror.

There is practically no end to this sort of thing, for if we
have a scientist at our side he will show us the mirror in
a series of experiences that we will call atoms, electrons,
protons. I don't mean that the cause of these experiences
is within us, but merely that these new words mark new
experiences that we are having with what at first we called
a star, then a candle, and now a mirror. They are perfectly
real—though they appear to have been capable of totally
confusing Professor A. S. Eddington, who in the Introduction to *The Nature of the Physical World*, speaks of
what he calls his "scientific" experiences with his library
table as if they were of a totally different order from his
ordinary experiences with the same innocent object of furniture. I wouldn't say that his reasoning was wrong, but
merely that he was moved by his desire to tell a fairy story
into forgetting what he was talking about. I don't believe
that "reasoning" can ever be wrong, or that a human "rea-

sone" can ever be quite right. To arrive at a perfect conclusion he would have to have a perfect will and a perfect experience. Wrong reasoning is unreason and even as to that, Reason itself is the sole judge. What we call wrong reasoning is a wrong conclusion logically deduced from wrong premises which we honestly (or dishonestly) take for truth.

A schoolboy in a class in logic, may recite: "All men are mortal. Socrates was a man. Therefore, all men are Socrates." He seems to have lost his reason, but not even lunatics really do that. They merely lose their wits, their power of receiving evidence. In all probability the boy was merely trying to remember what to him was a meaningless formula, and was a poor parrot rather than a poor philosopher.

But let us suppose that he was really thinking. What takes place is this. He slips into his argument some assumption not justified by the words' conventional meaning. He is saying to himself: "All men are mortal, even Socrates. I am a man, or almost one, therefore I am mortal and resemble Socrates—in this respect if no other." The word "Socrates" has spread out so as to include the meaning "mortal" in general. The boy's actual syllogism was: "All men are mortal. I am a man. Therefore I am mortal." Perfectly logical. Thus we say that women are illogical. They are not, they are merely given to attaching their own meanings to words, as suits their fancy.

This slipping in of unexpressed assumptions does not always lead to wrong conclusions, for the unexpressed assumption may be true, humanly speaking. The conclusion is never illogical, if we understand the words as really *meant*. But the assumption clearly may be wrong, whether expressed or not—as when we assumed, after seeing a point of brightness, that we then had all the experiences that that point of brightness could give us, or that we knew what they would be if we were to have them. Probably we should have admitted that such was not an assumption warranted by past experiences, but we forgot to take our limitations into consideration. We are always forgetting relevant factors like this.

We make this mistake if we assume that, at the present point of scientific inquiry, we have in hand all the experiences either a star or a candle can give us; that these flashes in the microscope and these instrument-readings that so very few of us have individually read which constitute our experience with electrons, mark the end. We even go beyond this, and assume that mere theories are final, though they may be but guesses at future experience, or at experiences that we may never hope to have. All such theories are based upon the assumption that we have exhausted the possibilities of some source of experience, have in fact already as good as experienced the unexperienced, which therefore can contain no surprises—that is, such is the assumption if an unverified theory is regarded as a fact. The perfect scientist will remember that his assumptions are tentative, that he is reasoning from false premises. But who is a perfect scientist?

There is one instance when our power of inference really does seem to be warped, and that is when we are blinded by partiality, or prejudice, or passion, when the will itself moves against the purity of thought. But here again the twist is not in deduction. Under the influence of passion, we are especially prone to forget circumstances that tell against the conclusion that we wish to arrive at. We make all sorts of assumptions that are not warranted. We treat falsehoods as facts of experience, and refuse to see facts that we do not like.

This brings us to the gist of the whole matter, the part played by the will in the reasoning process. When I said in one of the earlier chapters that before we reason we have to choose, I may have been accused by some readers of begging the question, of reducing all reason to a process of making excuses, of justifying a position already taken up by prejudice. This is precisely what takes place if our will is a will to establish as true some particular and wished-for conclusion—and it is notorious that this wish becomes emphatic in proportion to our consciousness of lack of knowledge, so that over-emphasis is an almost invariable sign of a guilty conscience.

But though now it is the will that is unreasonable, it gets in its deadly work not by warping our deductions but our judgment, our power of induction, of receiving, weighing, and measuring. Induction seeks to arrive at a general rule or law or principle by discovering some resemblance that runs through a series of experiences. We did this when we said, "All men are mortal." We merely meant our experiences with mankind showed what we call death coming to every individual after a certain limit of age. We accepted a great deal of hearsay. We assumed that there was no man now living who was above the limit. What is more, we assumed that no man now living would go on living forever, though this was pure theory, and based upon nobody's actual experience.

In order to give this theory any foundation, we had to assume that there is a certain constancy at the heart of things, that the past is some indication of the future. But as the future is never seen quite to repeat the past, and as we could not be certain that it would go on doing so in any case, our premise was not altogether justified. And so our conclusion, though logical, was not absolutely sure upon the data indicated. We should have said, If all men are mortal, If Socrates was a man, *then* Socrates was mortal. But we were justified, apart from the sage's conspicuous absence from the present scene, in concluding that Socrates was *probably* mortal, and long since dead—for probability is nothing but a word used to indicate that, of a given list of instances, all, or at least the majority, point in one direction. Probability never reaches certainty unless all the instances are certain, or unless it can be said that all relevant instances were certainly included in the list, or unless it is certain that the list is closed.

Even to give to probability that likelihood which is the essence of what we mean by it, it is necessary to believe in some degree of constancy in nature. Bertrand Russell is unable to justify any such belief, judging from all that he says, but most of the rest of us have been able to do so by observing in nature what may be called habits. (This does not mean that nature is controlled by its habits, since it is obviously controlled by what made the habits.) So we say that it is unlucky to bet on stocks, because stocks have a habit of falling immediately after one buys them, or of going up immediately after one sells them short. But (widespread experiences of last fall to the contrary notwithstanding) this does not imply that one can't, by some fluke, make a good speculation—more especially if one is deaf to tips. The future has always a blessed uncertainty and will have a degree of novelty, or it wouldn't be the future.

But the will I had in mind when I said that will was necessary to reason was that will which is willing to learn the truth at whatever cost, and is anxious only for that. It is often said that nobody is willingly deprived of the truth. It might better be said that nobody is deprived of the truth in any other way. Not that we ever learn the whole truth about anything. No will is even so perfect as to permit recognition of all the truth that is known. We are fond of admitting that we sometimes deceive ourselves. But I wonder if we ever do? We almost do. But there is something in us that knows when we are led aside by desire—by which I mean the desire for some particular thing, or conclusion, rather than that high desire for whatever may truly be. Of course when we leap in imagination ahead of the next step we can deceive ourselves, or could if we did not know that we were leaping; for then we enter the realm of pure ignorance—in so far as the ahead is really ahead. But we know perfectly well what we actually do see—I mean we know what we are actually experiencing. Before we can even shut our eyes to it, it is too late not to have known.

So I am of the opinion that an ill will, which desires not to know but rather to excuse itself, does so not by influencing deduction, but by tampering with the premises and by the introduction of assumptions that are unwarrantable and that we know to be unwarrantable. Though we may still make mistakes in our conclusions, even when we willingly use all the sight we have, these are not moral mistakes nor strictly speaking logical ones, but are merely that admixture of human error due to lack of sight—i. e., of information, and therefore not essentially ours but our ignorance's.

Faith, then, is not deceived, because there is no denying the experiential quality of experience nor the reasonableness of reason. Right Reason is that which springs from

good will, without bias, and to be entirely right in its results would have to have a perfect eye. Truth is not relative, but our knowledge most emphatically is, otherwise
our whims would themselves become absolutely true. It
is none the less true that our own individual truths are true
for us, in the sense that they are the best we can get, if it
be also true that our will is good.

Nor would it be well for me, I think, to attempt to live
too far ahead of my own knowledge, depending upon the
knowledge, real or professed, of others. I might come to
behave as if I had that other knowledge, and turn into a
prig. On the other hand, it would not be well for me to
avoid what I believe to be likely sources of further information, for fear of learning something that I do not
wish to know. And this raises the great question of authority.

One is certain to meet with *some* authority, with what
purports to be information beyond one's immediate grasp.
In choosing a teacher one can but use honestly such information as one already has. The desire to encounter influence
that one trusts will be untrustworthy manifests itself as
what is known as temptation. There is no mistaking it.
Nobody can be honestly mistaken, for then the mistake is
not his. But we all may be honestly uninformed, or misin-
formed—quite possibly by dishonest misinformers. May
their sins be upon their own heads!

E. THINKING THE UNTHINKABLE

Santayana's confusion comes from two things: (a) an
attempt to describe the Absolute, the Uncreate, in positive, comprehensible terms, and so reduce it to ourselves
—as if we could actually stand behind creation and move

about before creation's dawn; and (b) an attempt to adhere to the Modernist habit of describing cause in terms that are lower than its effect, thus making cause seem to proceed from effect—as when he speaks of habit as the governor of material life. Not only does he here contradict his previous statement that matter is the cause of all things, and hence ungoverned; but he makes habit, which is but a manifestation of some power whose acts have a family resemblance—he makes habit a power in itself, as if it created the power of which it is a manifestation.

F. WANTED, A FIXTURE

People in general, and Bertrand Russell in particular, have failed to realize the utterly devastating results that would flow from universal instability, results that in their sum would abolish all result. Thus Russell in his *Philosophy* holds that such chaos has its limits; that the general theory of relativity permits us to form accurate images of events that are going on at not too great a distance; that there is only a margin of error, marked by the time that it takes light to go to the place of the event and return to us with its report.

But to suppose that we can measure this interval is to suppose that we have a measuring-rod of constant length and a clock with a constant speed, and this again supposes that the earth is at rest. For if the earth is in motion there will be variations in rods and clocks; the size of the earth itself, the basis of all measurement in astronomy, will be called in question. The same difficulty arises if we try to prove that the earth is moving about a fixed point. We can't know how fast we are moving unless we know the distance of this point; and we can't know the distance of this point unless we know how fast we are moving, which can only be ascertained if we are not moving. Russell attempts to meet the question fairly and squarely, but at critical moments his hand falters, and he slips into his argument the unacknowledged assumption that at some time in our experiments we have a fixed earth at our disposal.

Thus he says that it makes no difference in our calculations whether the earth or the sun be at rest. It would make no difference if someone would present us with a reliable clock and a reliable measuring-rod—gifts that would remain reliable only upon a veritable *terra firma*. Nor will the supposition that the physical universe is limited furnish us with a practical point of departure. To say that the universe as a whole is standing still is to say that it has a center of gravity, or something of the sort, which is standing still. To make use of this it would be necessary to identify and locate it. Nor is there any meaning in saying that this point is standing still unless we admit the existence of something else, which in its very nature is moveless. I offer no solution of the difficulty with which physics is at present confronted, though I have little doubt that some constant will eventually be discovered, or at least elected, by science. Meanwhile it is obvious that we must believe in this constant in order to believe in the existence of anything, even ourselves—a belief which I, for one, find it unmannerly to abandon.

For the moment we are forced by Relativity Theory to the apparently ridiculous assumption that Galileo's (somewhat apocryphal) muttered remark, "The earth do move!" was uttered in haste. There is difficulty even in admitting that the earth turns on its axis, for this motion also would play havoc with our instruments unless its amount could be ascertained, and to ascertain it we would have to know our exact distance from the center in terms of a rod located at the center—not all of which rod could be at the center! We might assume that the rotation is constant, and even find some proof of it, and this would give us the basis for measuring relative distances—which is all we perhaps have a right to hope for. To deny the earth's rotation would bring another difficulty—that of accounting for the apparent revolution of the starry heavens every twenty-four hours. If the earth does not rotate, the motion of the heavens is real, and if our estimate of their distance is correct this would require them to move far faster than light. But at the speed of light, any physical mass is reduced to a point. This theory of the reduction of mass, itself destroys the physical existence of a light-ray regarded as a particle. And if it is not a particle what is it? Certainly it exerts a force when it strikes. Talk about mysticism! Philosophical mysticism is plain sailing compared with present-day science. Yet Dr. Watson tells us that there is no mystery even in the making and development of a man!